JAMES NESTOR has written for publications including *Outside*
*Scientific American*, *The New York Times*, *BBC*, *Men's Journal*
An inveterate adventurer, Nestor joined a doomed surfing exp
Norway and Russia, in which he and his team became the fir
the breaks of the Arctic Circle. He runs his 1978 Mercedes 300L
cooking oil and lives in a house he rebuilt himself in San Francis
at *mrjamesnestor.com*.

# PRAISE FOR *DEEP*

"Freediving, the sport that harnesses the mammalian dive reflex to survive deep plunges, can be a boon for marine researchers, avers James
Nestor. We meet a salty cast of them, such as the 'aquanauts' of Aquarius,
a marine analogue of the International Space Station submerged off the
Florida Keys. Equally mesmeric are Nestor's own adventures, whether
spotting bioluminescent species from a submarine in the bathypelagic
zone, or freediving himself—and voyaging into humanity's amphibious
origins in the process." —*Nature*

"The deeper the book ventures into the ocean, the more dramatic and
unusual the organisms therein and the people who observe them . . .
Through [Nestor's] eyes and his stories, it's a journey well worth taking."
—David Epstein, *New York Times Book Review*

"Nestor is crisp with his fun, seafaring facts; he is sober with his sprin-
kling of environmental bulletins. The book never preaches, and it's a
zippy read." —*Los Angeles Times*

"One of the best journalistic-style diving books I've read since Tim
Ecott's classic, *Neutral Buoyancy* . . . James Nestor is a fine observer."
—*Diver Magazine*

"*Deep* is a beguiling and profoundly informative book that reminds us, on
every page, that our Earth is still awash in mystery and wonder."
—Jon Mooallem, author of *Wild Ones*

# DEEP

JAMES NESTOR

# DEEP

FREEDIVING, RENEGADE
SCIENCE, AND WHAT THE OCEAN
TELLS US ABOUT OURSELVES

PROFILE BOOKS

This paperback edition published in 2015

First published in Great Britain in 2014 by
PROFILE BOOKS LTD
3 Holford Yard
Bevin Way
London WC1X 9HD
www.profilebooks.com

First published in the United States of America in 2014 by
Houghton Mifflin Harcourt

Portions of this book originally appeared in
slightly different form in *Outside* and *Men's Journal*

THIS BOOK PRESENTS THE IDEAS AND EXPERIENCES OF ITS
AUTHOR. IT IS NOT INTENDED TO PROVIDE INSTRUCTION FOR
FREEDIVING. THE AUTHOR AND THE PUBLISHER DISCLAIM
LIABILITY FOR ANY ADVERSE EFFECTS RESULTING DIRECTLY OR
INDIRECTLY FROM INFORMATION CONTAINED HEREIN.

5 7 9 10 8 6

Printed and bound in Great Britain by
CMI Group (UK) Ltd, Croydon CR0 4YY

A CIP catalogue record for this book is available from the British Library.

ISBN 978 1 78125 066 2
eISBN 978 1 84765 906 4

MIX
Paper | Supporting
responsible forestry
FSC® C171272

# Contents

| 0 | 1 |
| -60 | 12 |
| -300 | 27 |
| -650 | 54 |
| -800 | 80 |
| -1,000 | 98 |
| -2,500 | 124 |
| -10,000 | 157 |
| -35,850 | 198 |

| Ascents | 221 |
| Epilogue | 227 |
| Acknowledgments | 233 |
| Notes | 239 |
| Bibliography | 251 |
| Index | 259 |

# 0

I'M A GUEST HERE, a journalist covering a sporting event that few people have heard of: the world freediving championship. I'm sitting at a cramped desk in a seaside hotel room that overlooks a boardwalk in the resort town of Kalamata, Greece. The hotel is old and shows it in the cobweb cracks along the walls, threadbare carpet, and dirt shadows of framed pictures that once hung in dim hallways.

I've been sent here by *Outside* magazine, because the 2011 Individual Depth World Championship is a milestone for competitive freediving—the largest gathering of athletes in the history of the little-known sport. Since I've lived my whole life by the ocean, still spend much of my free time in it, and often write about it, my editor thought I'd be a good fit for the assignment. What he didn't know was that I had only a superficial understanding of freediving. I hadn't done it, didn't know anyone who had, and had never seen it before.

I spend my first day in Kalamata reading up on the competition rules and the sport's rising stars. I'm not impressed. I Google through photographs of competitive freedivers in mermaid outfits, flashing hang-loose signs while floating upside down in the water, and blowing intricate bubble rings from the bottom of a swimming pool. It seems like the kind of oddball hobby people take up, like badminton or Charleston dancing, so they can talk about it at cocktail parties and refer to it in their e-mail handles.

Nonetheless, I have a job to do. At five thirty the following morning, I'm at the Kalamata marina talking my way onto a twenty-seven-foot sailboat that belongs to a scruffy Québecois expat. His is the only spectator boat allowed out at the competition, which is held in the deep open waters about ten miles from the Kalamata marina. I'm the only journalist aboard. By 8:00 a.m., we've tied up to a flotilla of motorboats, platforms, and gear that serves as the competitors' jumping-off point. The divers in the first group arrive and take positions around three yellow ropes dangling off a nearby platform. An official counts down from ten. The competition begins.

What I see next will confound and terrify me.

I watch as a pencil-thin New Zealander named William Trubridge swallows a breath, upends his body, and kicks with bare feet into the crystalline water below. Trubridge struggles through the first ten feet, heaving broad strokes. Then, at about twenty feet, his body loosens, he places his arms by his sides in a skydiver pose, and he sinks steadily deeper until he vanishes. An official watching a sonar screen at the surface follows his descent, ticking off distances as he goes down: "Thirty meters . . . forty meters . . . fifty meters."

Trubridge reaches the end of the rope, some three hundred feet down, turns around, and swims back toward the surface. Three agonizing minutes later, his tiny figure rematerializes in the deep water, like a headlight cutting through fog. He pops his head up at the surface, exhales, takes a breath, flashes an okay sign to an official, then gets out of the way to make room for the next competitor. Trubridge just dove thirty stories down and back, all on one lungful of air—no scuba gear, air tube, protective vest, or even swim fins to assist him.

The pressure at three hundred feet down is more than nine

times that of the surface, strong enough to crush a Coke can. At thirty feet, the lungs collapse to half their normal size; at three hundred feet, they shrink to the size of two baseballs. And yet Trubridge and most of the other freedivers I watch on the first day resurface unscathed. The dives don't look forced either, but natural, as if they all really belong down there. As if we all do.

I'm so dazzled by what I see that I need to tell somebody immediately. I call my mother in Southern California. She doesn't believe me. "It's impossible," she says. After we talk about it, she dials some friends of hers who've been avid scuba divers for forty years and then calls me back. "There is an oxygen tank at the seafloor or something," she says. "And I suggest you do your research before publishing any of this."

But there was no oxygen tank at the end of the rope, and if there had been, and if Trubridge and the other divers had actually breathed some of it before ascending, their lungs would have exploded when the air from the tank expanded in the shallower depths, and their blood would have bubbled with nitrogen before they reached the surface. They'd die. The human body can withstand the pressures of a fast three-hundred-foot underwater ascent only in its natural state.

Some humans handle it better than others.

Over the next four days, I watch several more competitors attempt dives to around three hundred feet. Many can't make it and turn back. They resurface with blood running down their faces from their noses, unconscious, or in cardiac arrest. The competition just keeps going on. And, somehow, this sport is legal.

For most of this group, attempting to dive deeper than anyone — even scientists — ever thought possible is worth the risk of paralysis or death. But not for all of them.

I meet a number of competitors who approach freediving with a more sane outlook. They aren't interested in the face-off with mortality. They don't care about breaking records or beating the other guy. They freedive because it's the most direct and intimate way to connect with the ocean. During that three minutes beneath the surface (the average time it takes to dive a few hundred feet), the body bears only a passing resemblance to its terrestrial form and function. The ocean changes us physically, and psychically.

In a world of seven billion people, where every inch of land has been mapped, much of it developed, and too much of it destroyed, the sea remains the final unseen, untouched, and undiscovered wilderness, the planet's last great frontier. There are no mobile phones down there, no e-mails, no tweeting, no twerking, no car keys to lose, no terrorist threats, no birthdays to forget, no penalties for late credit card payments, and no dog shit to step in before a job interview. All the stress, noise, and distractions of life are left at the surface. The ocean is the last truly quiet place on Earth.

These more philosophical freedivers get a glassy look in their eyes when they describe their experiences; it's the same look one sees in the eyes of Buddhist monks or emergency room patients who have died and then been resuscitated minutes later. Those who have made it to the other side. And best of all, the divers will tell you, "It's open to everyone."

Literally everyone — no matter your weight, height, gender, or ethnicity. The competitors gathered in Greece aren't all the toned, superhuman Ryan Lochte–type swimmers you might expect. There are a few impressive physical specimens, like Trubridge, but also chubby American men, tiny Russian women, thick-necked Germans, and wispy Venezuelans.

Freediving flies in the face of everything I know about surviving in the ocean; you turn your back on the surface, swim away from your only source of air, and seek out the cold, pain, and danger of deep waters. Sometimes you pass out. Sometimes you bleed out of your nose and mouth. Sometimes you don't make it back alive. Other than BASE jumping — parachuting off buildings, antennas, spans (bridges), and earth (geological formations) — freediving is the most dangerous adventure sport in the world. Dozens, perhaps hundreds, of divers are injured or die every year. It seems like a death wish.

And yet, days later, after I return home to San Francisco, I can't stop thinking about it.

I BEGIN TO RESEARCH FREEDIVING and the claims made by competitors about the human body's amphibious reflexes. What I find — what my mother would never believe and what most people would doubt — is that this phenomenon is real, and it has a name. Scientists call it the mammalian dive reflex or, more lyrically, the Master Switch of Life, and they've been researching it for the past fifty years.

The term *Master Switch of Life* was coined by physiologist Per Scholander in 1963. It refers to a variety of physiological reflexes in the brain, lungs, and heart, among other organs, that are triggered the second we put our faces in water. The deeper we dive, the more pronounced the reflexes become, eventually spurring a physical transformation that protects our organs from imploding under the immense underwater pressure and turns us into efficient deep-sea-diving animals. Freedivers can anticipate these switches and exploit them to dive deeper and longer.

Ancient cultures knew all about the Master Switch and employed it for centuries to harvest sponges, pearls, coral, and food hundreds of feet below the surface of the ocean. European visitors to the Caribbean, Middle East, Indian Ocean, and South Pacific in the seventeenth century reported seeing locals dive down more than one hundred feet and stay there for up to fifteen minutes on a single breath. But most of these reports are hundreds of years old, and whatever secret knowledge of deep diving these cultures harbored has been lost to the ages.

I begin to wonder: If we've forgotten an ability as profound as deep diving, what other reflexes and skills have we lost?

I SPENT THE NEXT YEAR and a half looking for answers, traveling from Puerto Rico to Japan, Sri Lanka to Honduras. I watched people dive to one hundred feet and spear satellite transmitters onto the dorsal fins of man-eating sharks. I rode thousands of feet

5

down in a homemade submarine to commune with luminous jellyfish. I talked to dolphins. Whales talked to me. I swam eye-to-eye with the world's largest predator. I stood wet and half naked inside an underwater bunker with a group of researchers strung out on nitrogen. I floated in zero gravity. I got seasick. And sunburned. And a really sore back from flying tens of thousands of miles in coach. What did I find?

I discovered that we're more closely connected to the ocean than most people would suspect. We're born of the ocean. Each of us begins life floating in amniotic fluid that has almost the same makeup as ocean water. Our earliest characteristics are fishlike. The month-old embryo grows fins first, not feet; it is one misfiring gene away from developing fins instead of hands. At the fifth week of a fetus's development, its heart has two chambers, a characteristic shared by fish.

Human blood has a chemical composition startlingly similar to seawater. An infant will reflexively breaststroke when placed underwater and can comfortably hold his breath for about forty seconds, longer than many adults. We lose this ability only when we learn how to walk.

As we grow older, we develop amphibious reflexes that enable us to dive to incredible depths. The stresses of these depths would injure or kill us on land. But not in the ocean. The ocean is a different world with different rules. It's a place that often requires a different mindset to comprehend.

And the deeper we go in it, the weirder it gets.

When you're in the first few hundred feet, the human connection to the ocean is physical—you can taste it in your salty blood, see it in the gill-like slits of an eight-week-old fetus, and sense it in the amphibious reflexes humans share with marine mammals.

Past the limit where the human body can freedive and survive, about seven hundred feet, the connection to the ocean becomes sensory. You can see it reflected in deep-diving animals.

To survive in this lightless, cold, and pressurized environment, animals such as sharks, dolphins, and whales have developed extra senses to navigate, communicate, and see. We too share these extrasensory abilities; like the Master Switch, they are remnants of our collective past in the ocean. These senses and reflexes are latent and mostly unused in humans, but they have not disappeared. And they seem to revive when we desperately need them.

It's this connection—between the ocean and us, between us and the sea creatures with whom we share a great deal of DNA—that drew me deeper and deeper still.

AT SEA LEVEL, WE ARE ourselves. Blood flows from the heart to the organs and extremities. The lungs take in air and expel carbon dioxide. Synapses in the brain fire at a frequency of around eight cycles per second. The heart pumps between sixty and a hundred times per minute. We see, touch, feel, taste, and smell. Our bodies are acclimatized to living here, at or above the water's surface.

At sixty feet down, we are not quite ourselves. The heart beats at half its normal rate. Blood starts rushing from the extremities toward the more critical areas of the body's core. The lungs shrink to a third of their usual size. The senses numb, and synapses slow. The brain enters a heavily meditative state. Most humans can make it to this depth and feel these changes within their bodies. Some choose to dive deeper.

At three hundred feet, we are profoundly changed. The pressure at these depths is nine times that of the surface. The organs collapse. The heart beats at a quarter of its normal rate, slower than the rate of a person in a coma. Senses disappear. The brain enters a dream state.

At six hundred feet down, the ocean's pressure—some eighteen times that of the surface—is too extreme for most human bodies to withstand. Few freedivers have ever attempted dives to this depth; fewer have survived. Where humans can't go, other animals can. Sharks, which can dive below six hundred and fifty feet, and much deeper, rely on senses beyond the ones we know. Among them is magnetoreception, an attunement to the magnetic pulses of the Earth's molten core. Research suggests that humans have this ability and likely used it to navigate across the oceans and trackless deserts for thousands of years.

Eight hundred feet down appears to be the absolute limit of the human body. Still, an Austrian freediver is willing to risk paralysis and death to go beyond that depth.

At a thousand feet down, the waters are colder and there's almost no light. Another sense clicks on: animals perceive their environment not by looking but by listening. With this extra sense, called echolocation, dolphins and other marine mammals can "see" well enough to locate a metal pellet the size of a rice grain from a distance of 230 feet, and they can distinguish between a Ping-Pong ball and a golf ball from 300 feet away. On land, a group of blind people have tapped into the ability to echolocate and use it to ride bikes through busy city streets, jog through forests, and perceive a building from a thousand feet away. This group isn't special; with the right training, we all can see without opening our eyes.

At twenty-five hundred feet below, the water is permanently black, and pressures are seventy-five times that of the surface. For the animals living at these depths, danger lurks in all directions. Electric rays have adapted by harnessing impulses inside their bodies to fatally shock prey and fend off predators. Scientists have discovered that every cell in the human body also contains an electrical charge. Tibetan Buddhist monks who practice the Bön tradition of Tum-mo meditation have learned to focus these cellular charges to warm their bodies during bitterly cold winters. Researchers in England have discovered that by controlling the output of cellular charges in our bodies, humans can not only create heat but treat many chronic diseases.

At ten thousand feet below, a black and unforgiving depth, we find sperm whales—whose behavior, surprisingly, more closely resembles our culture and intellect than any other creature's on the planet. Sperm whales may communicate with one another in ways that could be more complex than any form of human language.

At twenty thousand feet and below, the deepest waters harbor the world's most inhospitable environments. Pressures range from six hundred to a thousand times that of the surface; temperatures hover just above freezing. There is no light and very little food. And yet life persists there. These hellacious waters may in fact be the birthplace of all life on Earth.

TWO MILLION YEARS OF HUMAN history, two thousand years of science experiments, a few hundred years of deep-sea adventuring, one hundred thousand marine biology graduate students, countless PBS specials, Shark Week, and still, *still*, we've explored only a fraction of the ocean. Sure, humans have gone deep on occasion, but have they really seen anything? If you compare the ocean to a human body, the current exploration of the ocean is the equivalent of snapping a photograph of a finger to figure out how our bodies work. The liver, the stomach, the blood, the bones, the brain, the heart of the ocean – what's in it, how it functions, how we function within it – remain a secret, much of it hidden in the dark and sunless realms.

To be clear, this book has a downward trajectory. With each passing chapter, it will descend farther, from the surface to the bottom of the blackest sea. I'll go down as far as I physically can, then, for those depths I cannot access, I'll use a proxy – one of the many deep-diving animals with whom humans share unexpected and startling similarities.

The research and stories that follow cover only a sliver of the current research on the ocean and pertain specifically to the human connection within this realm. The scientists, adventurers, and athletes profiled here are only a handful of thousands of people now plumbing the sea's mysteries.

It's no coincidence that many of the researchers are freedivers. I learned early on that freediving was more than just a sport; it was also a quick and efficient way to access and research some of the ocean's most mysterious animals. Shark, dolphins, and whales, for instance, can dive a thousand feet or more, but there's no way of studying them at such depths. A handful of scientists have recently discovered that by waiting for these animals to come to the surface, where they feed and breathe, and then approaching them on their own terms — by freediving — they can study them far more closely than any scuba diver, robot, or sailor.

"Scuba diving is like driving a four-by-four through the woods with your windows up, air conditioning on, music blasting," one freediving researcher told me. "You're not only removed from the environment, you're disrupting it. Animals are scared of you. You're a menace!"

The more I immersed myself in this group, the more I wanted to share the close encounters they were having with their subjects. I began freediving on my own. I became a student of the form. I went deep.

And so, my freediving training is also a part of this book's downward spiral — a personal quest to overcome dry-land instincts (aka breathing), flip the Master Switch, and hone my body into a diving machine. Only by freediving could I get as close as physically possible to the animals who were teaching us so much about ourselves.

But freediving, I knew, had its limits. Even experienced divers usually can't go below 150 feet comfortably, and even when they do, they can't stay long. The average beginning freediver — me, for instance — isn't able to get past a few dozen feet for several frustrating months. To get to these deeper depths and see deep-sea animals that never come near the surface, I followed a different kind of freediver — a subculture of do-it-yourself oceanographers who are revolutionizing and democratizing access to the ocean. While other scientists working in government and academic institutions were filling out grant requests and reeling from funding cuts, these DIY researchers were building their own submarines out of plumbing parts, tracking man-eating sharks with iPhones, and cracking the secret language of cetaceans with contraptions made of pasta strainers, broomsticks, and a few off-the-shelf Go-Pro cameras.

To be fair, many institutions don't carry out this kind of research because they can't. What this group of DIY researchers was doing was dangerous — and often totally illegal. No university would ever permit its graduate students to motor miles out to sea in a beat-up boat and swim with sharks and sperm whales (which have eight-inch-long teeth and are the largest predators on Earth) or ride thousands of feet deep in an unlicensed and uninsured hand-built submarine. But these renegade researchers did it all the time, often on their own dimes. And with their slapped-together gear and shoestring budgets, they clocked more hours with the inhabitants of the ocean's depths than anyone before them.

"Jane Goodall didn't study apes from a plane," said one freelance cetacean-communication researcher working out of a lab he'd set up on the top floor of his wife's restaurant. "And so you can't expect to study the ocean and its animals from a classroom. You've got to get in there. You've got to get wet."

And so I did.

WHAT HOUSTON IS TO SPACE stations, a turquoise two-story tract home in Key Largo is to Aquarius, the world's only remaining underwater habitat. In front of the house, a mailbox is propped up on a cinder block and zip-tied to a pile of weathered wood. White gravel covers a driveway filled with grimy, decades-old cars. Go past a menacing chainlink fence and up a wooden staircase, and you'll find a sliding glass door that opens onto a room paneled in 1970s veneer. Mission Control is on the right.

Aquarius is run out of what's essentially a dorm room. There are oak cabinets in the hallway, threadbare sofas placed at odd angles in the living room, and sunburned guys in shorts and backward ball caps eating microwaved noodles in the kitchen.

Saul Rosser, operations director, invites me into the observation deck. Rosser, who is thirty-two and has worked at Aquarius for two years, is wearing a black polo shirt, baggy brown pants, white socks, and black shoes — the unofficial uniform of an engineer at leisure. In front of him on a sectional desk are three computer monitors, a red telephone, and a logbook. Rosser shakes my hand and then excuses himself. He needs to take a call.

"Ointment," a female voice crackles through the speaker.

"Copy on ointment," says Rosser.

"Applying ointment," says the voice.

"Copy on applying ointment," says Rosser.

A closed-circuit video feed in front of Rosser — one of ten displays on the computer monitors — shows a grainy image of a hand applying ointment to a knee.

"Ointment applied," says the voice.

"Copy on ointment applied," says Rosser.

Rosser documents every word by hand in the logbook. The speaker goes silent. He stares at the video screen and watches as the woman caps off the tube of ointment. A moment later, another video feed from a different angle shows the back of a woman as she walks across a tiny room and puts the ointment in a small white drawer. The video is pixelated, and it looks as though the transmission is coming from outer space. Except for the fact that the woman is young, blond, and wearing a bikini bottom and a T-shirt, which, in a way, makes Mission Control seem even more like a dorm room.

"Over," the woman's voice crackles through the speaker.

"Over," says Rosser.

The woman, Lindsey Deignan, is a sponge researcher from the University of North Carolina, Wilmington. She has been living inside Aquarius for eight days and won't surface for another two. She's got a scratch on her knee that needs medical attention and some healing time in the sun, but it won't get either anytime soon. There's no sun in Aquarius, and no doctor. Opening the back hatch and swimming straight to the surface would probably kill Deignan; her blood would boil and most likely shoot out of her eyes, ears, and other orifices.

In the name of science, Deignan and five other researchers, called aquanauts, have volunteered to have their bodies supercompressed to thirty-six pounds per square inch so they can dive for as long as they want without ever having to worry about decompression sickness. The only requirement is that once the aquanauts head down to Aquarius, which is located seven miles off the coast from where we're sitting, they'll have to stay there for a week and a half, until the mission is over. Then they'll be decompressed, a seventeen-hour process that brings their bodies back to surface pressure and allows the nitrogen gas to safely diffuse.

In the name of research, I've come here to see what these scientists get out of spending ten days living in the equivalent of a submerged Winnebago. Plus, I can't freedive yet, so this is the best way for me to sample the immersive approach to underwater research.

A doctor visiting Aquarius a few years back demonstrated what would happen to Deignan and the other aquanauts should they suddenly get claustrophobic and go AWOL without decompressing. He dove down and drew blood from an aquanaut who was just finishing a long mission, placed the blood in a vial, and then headed back to the surface. By the time the doctor reached the top, the blood was bubbling so violently it blew the rubber stopper off the vial.

"Imagine what would happen to your head," says Rosser, kicking his black comfort shoes from beneath the desk. Sissy Spacek in *Carrie* comes to mind.*

The prospect of bubbling blood is only one of the inconveniences of living underwater in a steel box. Even with air conditioners running on high, nothing ever really dries down there. This is why Aquarius aquanauts are usually half naked and why Deignan applied ointment to a tiny scratch on her knee. In the pervasive humidity, which ranges between 70 and 100 percent, infections are rampant. So is mold, and so are earaches. Some divers experience constant, hacking coughs.

* Symptoms of decompression sickness, which is caused by nitrogen coming out of the bloodstream and forming bubbles when pressure suddenly decreases, aren't always immediate. Studies with pigs and other animals show that nitrogen outgassing reaches critical levels about thirty minutes after an animal resurfaces after a deep dive. First the body's large joints, such as elbows, knees, and ankles, start throbbing. Skin becomes itchy and mottled. Limbs become paralyzed, and lungs feel as if they're burning. In extreme cases, death follows.

In 2007, Lloyd Godson, a twenty-nine-year-old from Australia, attempted to spend a month living just twelve feet underwater in a self-sustaining pod called Biosub. It wasn't the loneliness that eventually got to him — it was the wetness. Within a few days, the humidity inside Biosub was at 100 percent, water was dripping from the ceiling, and Godson's clothes were soaking wet and molding. He became dizzy, faint, panicky, and paranoid. He lasted less than two weeks. Crews in Aquarius have lived in similar conditions for up to seventeen days. Fabien Cousteau, the grandson of the famous French ocean explorer, is planning a thirty-one-day mission in Aquarius in 2014.

If the moisture in Aquarius doesn't get you, the pressure might. One hundred and twelve tons of water presses down on Aquarius at all times. To keep the water out, the habitat must be pressurized at a high level, which, at around sixty feet below the surface, works out to about two and a half times the pressure at sea level. Being inside Aquarius feels the opposite of what it would feel like to be thirteen thousand feet up. Bags of chips become pancake flat. Bread becomes dense and hard. Cooking facilities are limited to hot water and a microwave, and most food is vacuum-packed camping stuff. Years back, a surface-support-crew diver delivered a lemon meringue pie in an airtight container to the aquanauts. The pressure had flattened it into a thin sheet of white-and-yellow goo by the time it was opened.

ROSSER IS NOW WATCHING A video feed of aquanauts as they prepare for sleep. (He writes in the logbook that the aquanauts are preparing for sleep.) One checks the oxygen level on a back wall. (Rosser writes in the logbook that an aquanaut checked the oxygen level on a back wall.) This goes on for the next twenty minutes.

Aquarius is under twenty-four-hour surveillance. Microphones record conversations in every room. Each movement, motion, and action is logged. Air pressure, temperature, humidity, and carbon dioxide and oxygen levels are checked by a computer every few seconds. Valves are checked every hour. The smallest break in the system could have a domino effect that would lead to flooding in the living chamber, which would instantly drown the aquanauts. Rosser and the other managers are there to make sure it doesn't happen. So far, they've done a good job.

Over the past two decades, Aquarius has run more than 115 missions, and there's been only one death, and that was caused by a malfunction on a rebreather device and had nothing to do with the laboratory itself.

But the Aquarius team members have had their share of close calls. A generator caught fire during a hurricane in 1994, requiring aquanauts to evacuate immediately after decompressing into fifteen-foot-high waves. Four years later, in another storm with seventy-mile-an-hour winds, Aquarius was ripped from its foundation and almost destroyed. In 2005, the seas got so rough that Aquarius — which weighs 600,000 pounds — was dragged a dozen feet across the seafloor.

To the aquanauts, though, danger, close quarters, sleeping on wafer-thin bunk beds, eating flattened potato chips, and sitting around wet and semi-nude are a small price to pay to have unfettered access to the first six stories of ocean, a depth researchers call the photic zone.

LIFE IN THE FIRST FEW hundred feet of the sea is much like life on land, only there's a lot more of it. The ocean occupies 71 percent of the Earth's surface and is home to about 50 percent of its known creatures — the largest inhabited area found anywhere in the universe so far.

The depth of shallow waters, called the photic ("sunlight") zone, varies depending on conditions. In murky waters of bays near the mouths of rivers, it might extend down to only about forty feet or so; in clear, tropical waters, it can reach down to around six hundred feet.

Where there's light, there's life. The photic zone is the only place in the ocean where there's enough light to support photosynthesis. Although it makes up only 2 percent of the entire ocean, it houses around 90 percent of its known life. Fish, seals, crustaceans, and more all call the photic zone their home. Sea algae, which makes up 98 percent of the biomass in the ocean and can grow nowhere else but in the photic zone, is essential to all life on land and in the ocean. Seventy percent of the oxygen on Earth comes from ocean algae. Without it, we couldn't breathe.

How algae can generate so much oxygen and how that might be affected by climate change, nobody knows. That's part of what the aquanauts on Aquarius are trying to find out. They're also trying to crack more mystical marine riddles, like the secret behind coral's "telepathic" communication.

Every year on the same day, at the same hour, usually within the same minute, corals of the same species, although separated by thousands of miles, will suddenly spawn in perfect synchronicity. The dates and times vary from year to year for reasons that only the coral knows. Stranger still, while one species of coral spawns during one hour, another species right next to it waits for a different hour, or a different day, or a different week before spawning in synchronicity with its own species. Distance seems to have no effect; if you broke off a chunk of coral and placed it in a bucket beneath a sink in London, that chunk would, in most cases, spawn at the same time as other coral of the same species around the world.

The synchronous spawn is essential for corals' survival. Coral colonies must continuously expand outward to thrive. To remain healthy and strong, they must breed outside of their gene pool with neighboring colonies. Once released to the surface, the coral sperm and eggs have only about thirty minutes to fuse. Any longer, and the coral eggs and sperm will either dissipate or die off. Researchers have found that if the spawning is just fifteen minutes out of sync, coral colonies' chances of survival are greatly reduced.

Coral is the largest biological structure on the planet and covers some 175,000 square miles of the seafloor, and it can communicate in a way far more sophisticated than anyone ever thought.

And yet, coral is one of the most primitive animals on Earth. Coral has no eyes, no ears, and no brain.

There soon won't be much of it left. All over the world, coral colonies have been dying off at record rates. Fifty percent of the corals along Australia's Great Barrier Reef have died. In some areas of the Caribbean, such as Jamaica, coral populations have shrunk by over 95 percent. Colonies off the coast of Florida died off by 90 percent over the past decade. The causes are unclear but scientists believe pollution and climate change are to blame. In fifty years, coral may be completely gone, and disappearing with it will be one of the stranger unsolved mysteries in the natural world.

For the Aquarius aquanauts who were researching coral, their work is a race against time — one of many such races I'll encounter in the months ahead.

EVER SINCE ARISTOTLE PROPOSED turning a giant jar upside down, putting a man inside it, and sinking it, humans have devised all sorts of grand schemes to explore the shallow waters of the photic zone. Most of these either killed or maimed their occupants. The history of underwater exploration is paved with the bones of those who tried to go deep.

In the 1500s, Leonardo da Vinci drew up a sketch for a diving suit: it was made of pig leather, had a pouch at the chest to hold air, and a bottle at the waist to catch urine. (It was never built.) Years later, another Italian suggested putting a bucket with glass cutouts over his head and diving down twenty feet. (It failed in trials.) In the 1690s, an English astronomer named Edmond Halley, who would later have a comet named after him, proposed dropping a man inside an enormous wooden bucket and delivering air to him through wine barrels. (He never tried it.)

The first diving contraption capable of making it down to Aquarius's depth was invented around 1715 by John Lethbridge, a wool merchant who lived in Devon, England, with his seventeen children. The craft was constructed using a six-foot-long oak cylinder that had a glass porthole at its head and an armhole with a leather sleeve at each side. Air was fed through a hose at top. The whole thing looked extremely primitive and fragile but Lethbridge managed to take it down to around seventy feet for a half hour at a time—although, Lethbridge wrote, it was done "with great difficulty."

A century later, a Brooklyn machinist named Charles Condert debuted a more agile and "safe" means to explore the seafloor—the world's first self-contained underwater breathing apparatus, or scuba. The device consisted of four feet of copper tube, which was mounted onto Condert's back, and a pump made from a shotgun barrel, which pulled air into a rubber mask that covered Condert's face. Anytime Condert wanted to breathe, he'd just pump the gun-barrel contraption and receive a blast of fresh air. In 1832, Condert debuted the device in New York City's East River and became the world's first successful scuba diver. Later that day, when the copper tube broke off at twenty feet down, Condert became the world's first scuba fatality.

Other inventions soon followed. In England, John Deane attached a fireman's helmet to a rubber suit to create the first production dive suit. A pump on deck delivered air through a hose that was connected to the back of the helmet, allowing a diver, for the first time, to stay at depths of around eighty feet for about an hour. The Deane helmet was a great success, but it was dangerous. The compressed air pumped into the suit made it susceptible to sudden and extreme shifts of pressure during dives. If the helmet or air tube ruptured, the reversed pressure created a vacuum in the suit that "squeezed" the diver's body from inside out, forcing blood out of the nose, eyes, and ears. Squeezes became semiregular events. Some were so powerful that a diver's flesh would be ripped from his body. In one case, so much of a diver's body was torn away that there was nothing to bury but the helmet clogged with his bloody remains.

The deeper humans plunged into the ocean, the more grotesque and violent the consequences. In the 1840s, construction workers were using watertight structures called caissons to build underwater foundations for bridges and piers. To keep water out, the structures were filled with pressurized air from the surface. After being in them for just a few days, caisson workers usually reported maladies like rashes, mottled skin, difficulty breathing, seizures, and extreme joint pain. Then they began dying.

The condition became known as caisson disease or, more commonly, the bends, named for the excruciating pain that the afflicted workers felt in their knees and elbows. Scientists later discovered that the shift from pressurized air in the caissons to normal air at the surface was causing nitrogen gas to bubble in the workers' bodies and collect in their joints.

It would take another forty years for engineers to understand that it wasn't the deep water that was harming the ocean explorers — it was the deep-diving machines. Ironically, while Western divers in carefully constructed suits or caissons were drowning or getting their faces sucked off or suffering the bends at depths above sixty feet, two thousand miles to the south, Persian pearl divers were regularly plummeting to twice that depth and doing it with nothing more than a knife and a single breath of air. They suffered none of these maladies, and they had been diving to these depths for thousands of years.

Eventually, Western engineers developed elaborate systems to protect the body from underwater forces. They figured out how pressures change at depth and how oxygen can become toxic. Lethbridge's and Deane's primitive inventions eventually led to armored suits with compressed air, submarines, and scuba-diving decompression tables.

In 1960, Don Walsh, a U.S. Navy lieutenant, and Jacques Piccard, a Swiss engineer, took a steel chamber called *Trieste* down to 35,797 feet in the Pacific Ocean's Mariana Trench — the bottom of the deepest sea. Two years later, humans were living underwater.

The first underwater habitat, built by Jacques Cousteau, was set up thirty-three feet below the ocean's surface in an area off the coast of Marseilles. Called Conshelf, it was about as big as the cabin of a Volkswagen bus, and just as cold and wet. "The hazards are great and exceed the challenges," said Cousteau of Conshelf. In fact, the hazards were so great that Cousteau sent two underlings in his place. They lasted a week.

A year later Cousteau planted a more deluxe five-room model — with living room, shower, and sleeping quarters — on the seafloor off the coast of Sudan. Footage from the expedition, later featured in Cousteau's Oscar-winning documentary *A World Without Sun*, shows a kind of futuristic/French paradise where by day, aquanauts spent their time floating through Technicolor sea gardens, and by night, they smoked, drank wine, ate perfectly prepared French meals, and watched television. The aquanauts lasted a month. Their only complaint was the lack of women down there to "keep us company."*

By the late 1960s, more than fifty undersea habitats around the world were being built, and many more were planned. Australia, Japan, Germany, Canada, and Italy were all going deep. Cousteau predicted that future generations of humans would be born in underwater villages and "[adapt] to the environment so that no surgery will be necessary to permit them to live and breathe in water. It is then that we will have created the man-fish." The race for inner space, it appeared, was on.

And then it was off. After just a few years, all but a handful of the habitats were scrapped. Living underwater proved to be much more of a challenge, and far more expensive, than any-

---

* By the mid-1960s, the ocean floor was hot property, and deep-sea missions were getting increasingly bizarre and dangerous. Not to be outdone by the French, in 1965 the U.S. Navy placed a former Mercury 7 astronaut, Scott Carpenter, inside a 680-square-foot steel tube called SEALAB II and sunk it off the coast of La Jolla, California, to a depth of 203 feet. For an entire month, Carpenter lived in SEALAB II, testing equipment, receiving posts from a navy-trained bottle-nosed dolphin named Tuffy, and huffing a gas mixture of mostly helium. (If it didn't work, there was a chance Carpenter would suffer seizures, nausea, irreversible lung damage, or worse.) The experiment was a success, but the helium had a side effect: While sucking the gas in a decompression chamber after the dive, Carpenter could speak to his commanders only in a high-pitched, helium-distorted voice. The no-bullshit conversation between squeaking Carpenter and President Lyndon Johnson, who called to congratulate him on completing the mission, became legendary.

one had thought. Salt water ate away at metal structures; storms ripped foundations from the seafloor; aquanauts lived in constant fear of decompression sickness and infections.

This was the space age, after all; men were landing on the moon and building houses in orbit, so spending weeks underwater in a cold, wet box — in an environment you couldn't even see in, let alone *be* seen in — seemed pointless. And few land dwellers could relate to the research on microbiology and oxygen toxicity that was being conducted down there. Scientists had proved that humans could dive down to the deepest ocean floors and live underwater, but so what?

TODAY, ALMOST ALL OCEAN research is done topside via robots dropped from the decks of boats. Humans know more about the ocean's chemical composition, temperatures, and bathymetry (underwater topography), but they have also grown more physically and spiritually distanced from it.

Most marine researchers (at least, the ones I interviewed early on) never even get wet. Aquarius, one of the last oceanic institutions where researchers got wet and stayed wet for ten days at a time, was slated for closure.

I wanted to see it, this last piece of the institutional legacy of ocean exploration, before it joined the trash heap of inventions rusting away on the ocean floor. I wanted to see how the sanctioned experts researched the ocean before I headed out to spend a year with the renegades.

KEY LARGO, SEVEN MILES OUT, in hissing and storming seas. I am about to attempt my first scuba dive down sixty feet to Aquarius. I flash the captain of the motorboat that shuttled me here a thumbs-up, adjust the mouthpiece, and head down. I descend twenty, thirty, forty feet and notice a stream of bubbles belching from the seafloor, like an upside-down waterfall. An Aquarius safety diver stands wreathed in the bubbles, beckoning me closer. I kick toward him, duck my head, and, a few seconds later, reemerge in the air of the wet deck at the back of Aquarius.

"Please take off your wetsuit," says a man at the top of the metal staircase. He hands me a towel to put around my waist. "And welcome to Aquarius."

His name is Brad Peadro and he'll be leading my tour. Because even the tiniest puddle can take days or weeks to dry in Aquarius, all visitors are required to leave their scuba gear and wet clothes at the door. Clad in my towel, I follow Peadro through the deck and into a control room. The squawk of amplified voices from the PA and blasts of pressurized air echo against the steel walls. A few paces in, I see two men and two women sitting arm to arm around a kitchen table. They are marine biology graduate students from the University of North Carolina, Wilmington, and they're just finishing up a ten-day mission researching sponges and coral. Between them lies a flattened, half-empty bag of Oreos. "The long days do wear on you," says a pallid man named Stephen McMurray who is researching the population dynamics of sponges. He dips a spoon into a Styrofoam cup of instant noodles and looks through a window to the seafloor below.

"Nothing is ever dry down here," says John Hanmer, sitting across from him. "Ever." Hanmer, who is studying parrotfish, laughs and looks at his hands. Another aquanaut, Inga Conti-Jerpe, sits beside him. Her matted, frizzy hair clings to her scalp like wet plaster. "The pressure does interesting things to your skin," she says with a chuckle.

The aquanauts all laugh, then fall silent. They laugh again, then go silent again. I can't help but feel that everyone down here is a little off. Not in the cabin-fever kind of way that I expected; they are far too jolly for that. They seem, basically, drunk.

I learn that having your body pressurized to 36 psi for extended lengths of time can produce mild delirium. At higher pressures, more nitrogen dissolves in the bloodstream, eventually producing the same effect as nitrous oxide, or laughing gas. The more nitrogen in the bloodstream, the more whacked out the aquanauts feel. By the end of a ten-day mission, the whole group is on the equivalent of a Whip-It bender.

Lindsey Deignan, the aquanaut I watched apply ointment to

23

her knee from Mission Control the night before, looks especially dazed. "The longer we're down here, the larger the space seems," she says, smiling broadly. "It's now like triple the size. It's as big as a school bus! But it seems bigger than that!"

To me, the aquanauts' euphoric haze feels like an essential coping strategy in this dank, cramped, dangerous place. Moldy towels, rusting metal, and suffocating humidity are the main facts of life here. And you can't just get up and go home without having blood squirt out your eyes. To make matters worse, every thirty seconds or so, the crests and troughs of the waves at the surface shift the pressure inside Aquarius, requiring all of us to equalize our sinus cavities by popping our ears.

The tour continues. Peadro leads me three paces east, into the sleeping quarters — two rows of bunk beds stacked three high — and then back into the kitchen. The tour is over, he says. There's nothing else to see on Aquarius.

I noticed that we haven't seen a bathroom, and I ask Brad if we've passed it.

"We usually just go out the back there," he says, pointing to the wet deck entrance I just swam through. The front door of Aquarius doubles as its outhouse.

Toilets are notoriously difficult to manage in underwater habitats, mostly due to the constant shifts in air pressure, which can create vacuums inside the plumbing lines. In early underwater habitats, toilets would explode and splatter waste throughout the compartment. Aquarius's commode is an improvement, but it is so small and offers so little privacy that aquanauts prefer to do their business in the water out back. Even that has its problems. Sea life fights for the human "food." On one occasion, a male aquanaut who was submerged in the wet deck from the waist down had his ass bloodied by a hungry fish.

Peadro tells me to head back to the wet deck. At 36 psi, nitrogen usually takes ninety minutes to reach dangerous levels, but it can sometimes happen sooner; to be safe, Aquarius allows visitors a maximum of a half hour onboard. My time here is up.

I put on my wetsuit, splash through the door, and kick into the smoky blue water. The constant gurgle from my scuba regulator scares off everything around me; it's like I've gone bird watching with a leaf blower strapped to my back. And the wetsuit, tank, and knot of tubes around my body prevent me from even *feeling* the seawater.

Being inside Aquarius was the same way. Even though the habitat allows the aquanauts to do invaluable long-term research, sitting in that steel tube and looking at the ocean through windows and video screens was, to me, hopelessly isolating. I've felt far more connected to the ocean and its inhabitants surfing on its surface than sitting in a rubber-and-steel tube six stories beneath it.

BACK ON THE MOTORBOAT, I strip off my scuba gear and sit in the captain's cabin. Before I can leave, members of the Aquarius support crew need to dive down some canisters of food and supplies for the aquanauts.

The captain, an intense, sunburned man named Otto Rutten who has been working at Aquarius for more than twenty years, hands me a bottle of water. He tells me about some close calls he's had in this job—rescues in the high seas, explosions, emergency ascents.

"It was really the Wild West out here," he says. "I mean, we weren't even using scuba for a lot of the deliveries." He explains that scuba took too long and enabled him to make only a few dives at a time before the nitrogen gas in his bloodstream built up to dangerous levels. So instead, Rutten and the other crew members would just jump into bathing suits, put on fins and masks, and freedive the supplies down.

Swimming down there while carrying a bulky, airtight container and then coming back would take well over a minute. I mention to Rutten that he and the other divers must have stopped down at Aquarius to take a breath before returning to the surface. Rutten laughs and says that if he had, the high-pressure air would have probably killed him.

It was by stripping off all the gear—the tanks, weights, regulators, and buoyancy-control devices—that Rutten and his co-workers could dive deeper, more often, and four times as fast as someone wrapped in the most technologically advanced equipment.

I ask Rutten if he had any kind of special training to freedive to such depths.

"No, not really," he says. "It's easy. You just take a breath and go."

# −300

IN 1949, A STOCKY ITALIAN air force lieutenant named Raimondo Bucher decided to try a potentially deadly stunt off the coast of the island of Capri. Bucher would sail out to deep waters, take a breath, and go down one hundred feet to the bottom. Waiting there would be a man in a diving suit. Bucher would grab a baton with a parchment letter rolled inside it from the diver, then kick back up to the surface. If he completed the dive, he'd win a fifty-thousand-lira bet; if he didn't, he would drown.

Scientists warned Bucher that, according to Boyle's law, the dive would kill him. Formulated in the 1660s by the Anglo-Irish physicist Robert Boyle, this equation predicted the behavior of gases at various pressures, and it indicated that the pressure at a hundred feet would shrink Bucher's lungs to the point of collapse. He dove anyway, grabbed the baton, and returned to the surface smiling, with his lungs perfectly intact. He won the bet, but more important, he proved all the experts wrong. Boyle's law, which science had taken as gospel for three centuries, appeared to fall apart underwater.

Bucher's dive resonated with a long line of experiments — most of them very cruel and even monstrous by modern standards—that seemed to indicate that water might have life-lengthening effects on humans and other animals.

This line of inquiry arguably began in 1894, when Charles Richet rounded up several ducks and tied strings around their necks. He took half the group, tightened the strings until the birds couldn't breathe, then timed how long it took them to die. He then repeated the process with the other half, but these he strangled underwater. The ducks left in the open air lived only seven minutes, while the ducks kept underwater survived up to twenty-three. This was very odd. Both groups were deprived of oxygen in the same way, but the ducks put underwater lived three times longer.

Richet, who would eventually win a Nobel Prize for his work on the causes of allergic reactions, thought water might be affecting the ducks' vagus nerve. In both humans and ducks, this nerve extends from the brain stem to the chest and can slow the heart rate. Richet theorized that a slower heart rate would result in decreased use of oxygen and thus longer survival times.

He tested this theory by injecting one group of ducks with the drug atropine, which keeps the vagus nerve from slowing the heart rate. He left the second group untouched and atropine-free. He strangled both groups and timed how long it took them to die. They all died in about six minutes.

Then, with another group of ducks, he injected atropine and repeated his experiment, this time with the ducks underwater. The atropine-dosed ducks took more than twelve minutes to die underwater—twice as long as the ducks in open air. Even though the vagus nerve had been blocked with atropine and could not slow the heart rate, the water *still* had some inexplicable life-lengthening effect on the ducks. Richet took one atropine-drugged duck out of the water after twelve minutes, untied its neck, and resuscitated it. It lived.

Lung size, blood volume, and even the vagus nerve couldn't explain Richet's results. Water alone was extending their lives. He wondered if it had the same effect on humans.

In 1962, Per Scholander, a Swedish-born researcher working in the United States, confirmed that it did. He gathered a team of volunteers, covered them with electrodes to measure their heart rates, and poked them with needles to draw blood. Scholander had seen the biological functions of Weddell seals reverse in deep water; the seals, he wrote, actually seemed to *gain* oxygen the longer and deeper they dove. Scholander wondered if water could trigger this effect in humans.

He started the experiment by leading volunteers into an enormous water tank and monitoring their heart rates as they dove down to the bottom of the tank. Just as it had done with ducks, water triggered an immediate decrease in heart rate.

Next, Scholander told the volunteers to hold their breath, dive down, strap themselves into an array of fitness equipment submerged at the bottom of the tank, and do a short, vigorous workout. In all cases, no matter how hard the volunteers exercised, their heart rates *still* plummeted.

This discovery was as important as it was surprising. On land, exercise greatly increases heart rate. The volunteers' slower heart rates meant that they used less oxygen and therefore could stay underwater longer. This also explained, to some degree, why Bucher and those ill-fated ducks could survive up to three times longer in water than they could in open air: water had some powerful capacity to slow animals' hearts.

Scholander noticed something else: Once his volunteers were underwater, the blood in their bodies began flooding away from their limbs and toward their vital organs. He'd seen the same thing happen in deep-diving seals decades earlier; by shunting blood away from less important areas of the body, the seals were able to keep organs like the brain and heart oxygenated longer, extending the amount of time they could stay submerged. Immersion in water triggered the same mechanism in humans.

This shunting is called peripheral vasoconstriction, and it explains how Bucher could dive to below one hundred feet without suffering the lung-crushing effects that Boyle's law had predicted. At such depths, blood actually penetrated the cell walls of the organs to counteract the external pressure. When a diver descends to three hundred feet — a depth frequently reached by modern freedivers — vessels in the lungs engorge with blood, preventing them from collapse. And the deeper we dive, the stronger the peripheral vasoconstriction becomes.

Scholander found that a person need submerge only his face in water to activate these life-lengthening (and lifesaving) reflexes. Other researchers tried sticking a hand or a leg in the water in an attempt to trigger the reflex, but to no avail. One researcher even put volunteers into a compression chamber to see if pressure alone would trigger a similar diving reflex. No dice. Only water could trigger these reflexes, and the water had to be cooler than the surrounding air.

As it turns out, the tradition of splashing cold water on your face to refresh yourself isn't just an empty ritual; it provokes a *physical* change within us.

Scholander had documented one of the most extreme transformations ever discovered in the human body, a change that occurred only in water. He called it the Master Switch of Life.

Today, competitive freedivers are using the Master Switch to dive deeper and stay underwater longer than even modern scientists believe is possible.

On September 17, 2011, I traveled to Kalamata, Greece, to watch the modern-day masters of the Master Switch — one hundred of the world's best freedivers — test the absolute limits of our amphibious nature.

AT 7:00 P.M., THE OPENING ceremony of the Individual Depth World Championship is in full swing. Hundreds of competitors, coaches, and crew members from thirty-one countries are waving national flags and screaming their countries' anthems from an enormous stage built on a crowded boardwalk overlooking Kalamata Harbor. Behind them, a forty-piece marching band plays a ragged version of the *Rocky* theme as video highlights of freedivers plummeting three hundred feet are projected onto a thirty-foot screen. The whole scene looks like a low-rent Olympics.

Competitive freediving is a relatively new sport, and almost every year since Raimondo Bucher's hundred-foot dive in Capri — considered the first official competitive freedive — freedivers have been breaking records. The current world record for breath-holding underwater, set in 2009 by Frenchman Stéphane Mifsud, is eleven minutes, thirty-five seconds. In 2007, Herbert Nitsch, an Austrian freediver, dove down seven hundred feet on a weighted sled to claim a world record in absolute depth.

While nobody has ever drowned at an organized group freediving competition, enough freedivers have died outside of competition that it ranks as the second most dangerous adventure sport. The numbers are a bit murky; some deaths go unreported, and the statistics don't distinguish between deaths due to freediving alone and deaths due to freediving as part of other activities, like spearfishing. But one estimate of worldwide freediving-related fatalities over a three-year span revealed a nearly threefold increase: from 21 deaths in 2005 to 60 in 2008. Of the 10,000 active freedivers in the United States, about 20 will die every year, which works out to about 1 in 500. (In comparison, the fatality rate for BASE jumpers is 1 in 60; firefighters, about 1 in 45,000; and mountain climbers about 1 in 1,000,000.)

Just three months before the 2011 world championship, two deaths drew attention to the sport's dangers. Adel Abu Haliqa, a forty-year-old founding member of a freediving club in the United Arab Emirates, drowned in Santorini, Greece, during a 230-foot dive attempt. His body was never found. A month later, Patrick Musimu, a former world-record holder from Belgium, drowned while training alone in a pool in Brussels.

Competitive freedivers blame such deaths on carelessness, arguing that the fatalities are often associated with the divers going it alone or relying on machines for assistance — both very risky practices. "Competitive freediving is a safe sport. It's all very regulated, very controlled," said William Trubridge, the world-record freediver, when I talked to him before the opening ceremonies. "I would never do it if it wasn't." He pointed out that, during some 39,000 freedives over the previous twelve years, there had never been a fatality. Through events like the world championship, Trubridge and others hope to change freediving's dangerous image and bring it closer to the mainstream. Trubridge said he'd like to see it as an Olympic sport someday. The 2011 opening ceremony here in Greece, with all its blaring music and fast-edit videos, is meant to spread the word.

Onstage, the lights suddenly darken, the video screen dims, and the PA system goes silent. Moments later, strobe lights flash. The metronomic thump of an electronic bass drum pumps out of the speakers, joined soon after by canned handclaps and a bass riff that borrows heavily from "Another One Bites the Dust." Fireworks explode overhead. The freedivers cheer and dance around, waving national flags.

The freediving world championship is on.

FOR ALL ITS MAINSTREAM HOPES, competitive freediving has one glaring problem: It's almost impossible to watch. The playing field is underwater, there are no video feeds beamed back to land, and it's a logistical challenge to even get near the action. Today's staging area is a ragged twenty-foot-by-twenty-foot flotilla of boats, platforms, and air tanks; it looks like it was swiped from the set of *Waterworld*. To get there, I walk to the Kalamata marina and board a sailboat owned by a Québecois expat named Yanis Georgoulis. His is the only boat going to the competitions. Georgoulis tells me it will take about an hour to reach the flotilla. I use the time to further brush up on the complicated rules of today's competition.

The contest officially starts the night before a dive, when each competitor secretly submits the proposed depth of the next day's attempt to a panel of judges. It's basically a bid, and there's gamesmanship involved as each diver tries to guess what the other divers will do. "It's like playing poker," said Trubridge. "You are playing the other divers as much as you are playing yourself." The hope is that your foes will choose to do shallower dives than you can do or that they'll choose deeper dives than they can do and end up busting.

In freediving, you bust if you flub any one of dozens of technical requirements during and after the dive, or if you black out before you reach the surface, grounds for immediate disqualification. While not common in competitions (I'm told), blackouts happen often enough that layers of safety precautions are in place, including rescue divers who monitor each dive, sonar tracking from the flotilla, and a lanyard guide attached to each diver's ankle that keeps him or her from drifting off course — a potentially fatal hazard.

A few minutes before each dive, a metal plate covered in white Velcro is attached to a rope and sunk to the depth the competitor submitted the night before. An official counts down, and then the diver submerges and follows the rope to the plate, grabs one of the many tags affixed to it, and follows the rope back to the surface. About sixty feet under, the competitor is met by rescue divers who will assist him if he blacks out. If this blackout occurs so far down that the safety divers can't see him, the sonar will detect his lack of movement. The rope will then be hauled to the surface, dragging the freediver's body like a rag doll.

Divers who successfully resurface are put through a battery of tests known as the surface protocol. This regimen gauges the diver's coherence and motor skills by requiring him to, among other things, remove his facemask, quickly flash an okay sign to a judge, and say, "I'm okay." If you pass, you get a white card, validating the dive.

"The rules are there to make freediving safe, measurable, and comparable," said Carla Sue Hanson, the media spokesperson for the Association Internationale pour le Développement de l'Apnée (AIDA) or, as it's known in English, the International Association for the Development of Apnea, the freediving federation that has overseen the world championship since 1996. (*Apnea* is Greek for "without breathing.") "They are set up to ensure that, through the whole dive, the diver is in full control. That's what competitive freediving is all about: control."

As long as you're in control, it's all right if blood vessels burst in your nose and you come out looking like an Ultimate Fighter who took a beating. "The judges don't care how someone looks," Hanson said. "Blood? That's nothing. As far as the rules go, blood is okay."

After an hour, Georgoulis ties up to the flotilla. In the distance, a motorboat cuts a white line from the shore to deliver the first competitors to the site. Because of the extremely limited room on the flotilla and an adjacent motorboat, only judges, competitors, coaches, and a handful of staff are allowed at the event. There are no fans present. Luckily, I was able to talk my way onto Georgoulis's sailboat, which will be used as a makeshift locker room for contestants.

The first divers show up wearing hooded wetsuits and insectoid goggles, each moving with syrupy-slow steps as they warm up on the sailboat, staring with wide, lucid eyes. One, two, three — they slide into the sea like otters, then lie back looking semicomatose as their coaches slowly float them over to one of three lines dangling from the flotilla. A judge issues a one-minute warning, and the first competitor begins his descent.

Freediving is broken down into multiple disciplines. Today's is called constant weight without fins, or CNF. In CNF, a diver goes down using his lungs, body, and an optional weight that, if employed, must be brought back to the surface. Of the six categories in competitive freediving — from depth disciplines like free immersion (the diver can use the guide rope to propel himself up and down) to pool disciplines like static apnea (simple breath-holding) — CNF is considered the purest. Its reigning champion is Trubridge, who broke the world record in December 2010 with a 331-foot dive. Today he's trying for 305 feet, a conservative figure for him but the deepest attempt on the schedule. Before he arrives, a dozen other divers get things started.

An official on line one counts down from ten, announces, "Official top," and begins counting up: "One, two, three, four, five . . ." The countdowns let the divers know when to start gulping their last breaths of air and prepare to go deep. A female diver on line three, Junko Kitahama of Japan, has until thirty to go. She inhales a few final lungfuls, ducks her head beneath the water, and descends. As her body sinks, the monitoring official announces her depth every few seconds.

Two minutes later, a judge on the surface yells, "Blackout." Safety divers kick down along the rope and reemerge a half a minute later with Kitahama's body between them. Her face is pale blue, her mouth agape, her head craned back like a dead bird's. Through her swim mask, her wide eyes stare into the sun. She isn't breathing.

"Blow on her face!" yells a man swimming next to her. Another man grabs her head from behind and raises her chin out of the water. "Breathe!" he yells. Someone from the deck of a boat yells for oxygen. "Breathe!" the man repeats. But Kitahama doesn't breathe. She doesn't move.

A few agonizing seconds later, she coughs, jerks, twitches her shoulders, and flutters her lips. Her face softens as she comes to. "I was swimming and . . ." She laughs and continues. "Then I just started dreaming!" Two men slowly float her over to an oxygen tank sitting on a raft. While she recovers, another freediver takes her place and prepares to plunge even deeper.

Meanwhile, a diver on a different line takes one last breath, descends two hundred feet, touches down, and then, after three minutes, resurfaces. "Breathe!" his coach yells. He smiles, gulps, then breathes. His face is white. He tries to take off his goggles, but his hands are cramped and shaking. Lack of oxygen has sapped his muscle strength, and he just floats there, with blank eyes and a clownish grin.

Behind him another competitor resurfaces. "Breathe! Breathe!" a safety diver yells. The man's face is blue, and he isn't breathing. "Breathe!" another yells. Finally he coughs, jiggles his head, and makes a tiny squeaking sound like a dolphin.

For the next half an hour, divers come and go, and similar scenes play out. I stand in the sailboat with my stomach tightening, wondering if this is normal. All the competitors sign waivers acknowledging that heart attacks, blackouts, or drowning may be the price they pay to compete. But I have a feeling that competitive freediving's continued existence has a lot to do with the fact that the local authorities don't know what really goes on out here.

Trubridge arrives, wearing sunglasses and headphones, his arms appearing spidery next to his oversize torso. I can see his gargantuan lungs heaving even though I'm thirty feet away. He's so lost in a meditative haze that he looks half asleep as he enters the water, latches his ankle to the lanyard, and gets set to go.

A judge announces, "Official top," and a few seconds later Trubridge dives, kicking with bare feet, descending rapidly. The official announces, "Twenty meters," and I watch through the clear blue water as Trubridge places his arms at his sides and sinks effortlessly, drifting into the deep, and then is gone. The image is both beautiful and spooky. I try to hold my breath along with him and give up after thirty seconds.

Trubridge passes a hundred feet, a hundred and fifty feet, two hundred feet. Almost two minutes into the dive, the sonar-monitoring official announces, "Touchdown" — at 305 feet — and begins monitoring Trubridge's progress upward. After an agonizing three and a half minutes, Trubridge rematerializes. A few more strokes and he surfaces, exhales, removes his goggles, gives the okay sign, and says in his crisp New Zealand accent, "I'm okay." He looks slightly bored.

THE NEXT TWO DAYS ARE rest days. The courtyard at the Akti Taygetos Hotel buzzes with a dozen languages as teams gather around patio tables to sip bottled water, talk strategy, and e-mail worried relatives. The group here is largely male, mostly over thirty, and generally skinny. Some are short, a few are pudgy, and many have shaved heads and wear sleeveless T-shirts, action-strap Teva sandals, and baggy shorts. They hardly look like extreme athletes.

I find an empty table in the shade. I've scheduled an interview and a freediving lesson with Hanli Prinsloo, a national record holder from South Africa whom I met the previous day on Georgoulis's boat. She told me that for the past three months, she had been in Egypt training to break a world record, but she had come down with a sinus infection the previous week and had to pull out. She was now coaching friends, spreading good cheer, and patiently answering my many questions about the sport. She had also been urging me to try freediving myself.

So far, the mere thought of freediving made me claustrophobic. Aside from a few graceful and awe-inspiring dives from champions like Trubridge, most attempts looked awkward and dangerous. On the first day, seven competitors blacked out before reaching the surface; if they hadn't been rescued by the safety divers, they'd now be dead on the seafloor. The human body was no doubt uniquely equipped to dive deeper than I had ever imagined, but it still wasn't meant to descend to the depths these divers were attempting. It was just a matter of time before someone got hurt, or worse.

Prinsloo insisted that there was more to freediving than descending along ropes and trying to beat your opponents. "It offers a stillness," she told me on the boat, a kind of full-body meditation that could be found nowhere else. And there was no need to force yourself down to three hundred feet to find it. The most incredible transformation, she said, happened at around forty feet down. There, the force of gravity seemed to reverse; the water stopped buoying your body toward the surface and instead started pulling you deeper.

This was the "doorway to the deep," where everything changed, and anyone could pass through it — even me. To prove it, Prinsloo offered me an introductory, out-of-water session where we'd work on increasing my breath-holding capacity, the first step in learning to freedive. My personal breath-hold best was around fifty seconds; she promised that within two hours of training, I'd double it.

"WELL, HELLO!" PRINSLOO EXCLAIMS AS she approaches my poolside table. At thirty-four, she's tanned and fit, with long, dark brown hair; she actually looks like a natural athlete, unlike most of the freedivers I've seen. She grew up on a farm in Pretoria, South Africa, and spent her summers with her sister swimming in rivers and, she joked, speaking "a secret mermaid language." After discovering freediving in her twenties while living in Sweden, she moved back to South Africa. She now lives in Cape Town, where she runs the nonprofit conservation program I Am Water and works part-time as a motivational speaker and a yoga and freediving instructor.

We walk to a covered patio overlooking Messinian Bay and roll out yoga mats. The lesson begins with some basic poses to loosen the muscles around our chests. "If you could take your lungs out of your chest, they are completely flexible and you could blow them to whatever size," she says, then she puffs up her chest and exhales. What stops the lungs from expanding is the musculature around the ribs, chest, and back. Through stretching and breathing exercises, freedivers develop up to 75 percent more lung capacity than the average person. Nobody actually needs this extra capacity to start freediving, but, like a larger tank of gas, it can help you go deeper and stay under longer. Stéphane Mifsud, who set the world breath-hold record in 2009, boasts a 10.5-liter lung capacity; the average adult male's is 6 liters. Prinsloo can hold up to 6 liters of air in her lungs, compared to the average female, who can hold about 4.2.

Next, Prinsloo takes me through a few human-pretzel poses designed to help open up my lungs. While we're stretching, she explains how pressure works in water, and how it affects our lungs and bodies.

In the water, the deeper we go, the more the pressure increases and the more the air contracts. Seawater is eight hundred times denser than air, so diving down just ten feet causes the same change in air pressure as descending from an altitude of ten thousand feet to sea level. Anything with a flexible surface and air inside it — a basketball, a plastic soda bottle, human lungs — will be at half its original volume 33 feet underwater, a third of its original volume at 66 feet, a quarter at 99 feet, and so on.

When the basketball, plastic soda bottle, or pair of lungs returns to the surface, the air inside will quickly reinflate to its original volume. For freedivers, this plays hell on the body, especially the chest area. The breathing exercises and stretches Prinsloo is leading me through are meant to keep the chest muscles flexible so that if I start freediving, I'll be better able to handle these dramatic changes in volume and not black out or die.

We are now sitting cross-legged and facing each other, breathing into the three chambers of our lungs: the belly area, the sternum, and the top of the chest, just beneath the collarbones. Prinsloo says that most of us spend our lives breathing only at the very tops of our chests, meaning that we're accessing only part of our lungs. To store more oxygen for longer dives, I'll need to learn to breathe into the total volume of my lungs.

She directs me to draw a twenty-second breath into the belly area, sternum, and top of the chest. Doing this makes me feel nauseated, but I acclimate after a few minutes. Then Prinsloo pulls out her stopwatch and gets ready to time my first breath-hold attempt. I lie down on my mat, take one more enormous three-chambered breath and hold it. She starts the clock.

What feels like thirty seconds pass. I'm extremely nauseated. My head throbs. I imagine for a moment what it must be like to be a hundred feet underwater and feeling this awful. This thought triggers panic. A few seconds later, my body starts convulsing. I try to keep still but can't. Prinsloo stops the watch and tells me to exhale, then inhale. I sit up, shaking my head, feeling like a failure.

"Not bad," she says. "You've more than doubled your breath-holding on the first try." She shows me the stopwatch. I've just held my breath one minute and forty-five seconds.

I ask about the convulsions. She explains that the body responds to extreme breath-holding in three stages. Convulsions are the first-stage response. "You start reacting not from the lack of oxygen, but from the buildup of carbon dioxide," she says. "When that starts, it's just a caution that you've only got a few minutes to go before you *really* need to breathe." The second-stage response occurs when the spleen releases up to 15 percent more fresh, oxygen-rich blood into the bloodstream. This usually occurs only when the body goes into shock, an extreme state whose symptoms include low blood pressure, rapid heartbeat, and organ shutdown. But it also happens during extreme breath-holding. A freediver anticipates the spleen's delivery of fresh blood, feels it happen, and uses it as a turbo-charge to dive even deeper.

The third-stage response is the blackout, which happens when the brain senses that there's not enough oxygen for it to support itself and so shuts off, like a light switch, to conserve energy. Though the brain represents only about 2 percent of the body's weight, it uses 20 percent of the body's oxygen. The presence of liquid in the mouth or throat triggers another reflexive line of defense: the larynx automatically closes, stopping water from entering the lungs. Freedivers learn to sense the arrival of convulsions and spleen release, and they know exactly when to head back to the surface so the third-stage blackout won't occur. A freediver survives by understanding and respecting these mechanisms.

"There's a reason we're built with all these amazing rows of defense," Prinsloo says. "It's that we are meant to be underwater!" She shifts me into yet another yoga pose. "You are born to do this!"

I lie on my back for my final breath-holding attempt of the day. *Inhale, exhale, big inhale, hold.* Prinsloo starts the stopwatch. I close my eyes.

After what feels like about twenty seconds, I start gently convulsing again. I tell myself this is natural, to concentrate, keep relaxed, wait for the spleen to kick in. It's hard to wait. My chest feels pressurized and my heart pounds so forcibly that I can sense it in my hands, legs, crotch. I feel miserable.

"Stick with it, you can do this for so much longer. You're just at the first stage," Prinsloo reassures me. I stick with it. After what feels like ten more seconds, my stomach begins constricting, and my throat tenses. I feel claustrophobic. "Just a little longer . . . a little longer," she says gently. Soon my body feels electrified. I noticed I'm wriggling on the mat like a fish out of water. "Right now, your spleen is filling your body with fresh, oxygen-rich blood," she says. Moments later, I think I can sense what she's talking about. My body calms. The darkness of my closed eyes grows somehow darker; the ambient noise of the pool area fades; and I feel like I'm drifting off to . . .

"Breathe!" she says. I exhale, inhale, exhale. I'm dizzy, have trouble focusing through fluttering eyes, but I feel good. "How long do you think that was?" she asks me. I shrug and guess about a minute or so. She smiles. I didn't just double my breath-holding record during this lesson; I tripled it. The stopwatch reads three minutes, ten seconds.

HUMANS MAY WELL BE BORN to freedive, as Prinsloo insisted, but that doesn't mean it's easy. You still have to hold your breath a long time, exert yourself to your breaking point, and not freak out. I could now hold my breath for more than three minutes, but I hadn't tried diving any deeper than ten feet or so. And after what I'd seen, diving to even a few dozen feet was out of the question.

And yet, I was still determined to find out what it was like down there.

Three hundred feet is the halfway point to the photic zone. Even in the clearest oceans, the intensity of light at this depth is about .5 percent of what it is at the surface. With less light, there is less life than at shallower, brighter depths. The creatures who do live here must adapt to the twilight: fish have evolved large eyes to see better; sharks use electromagnetic senses to seek out prey; squids, microorganisms, and bacteria use a chemical process called bioluminescence to light their own way.

Getting down to this depth is arduous and often dangerous. Scuba divers can make it to three hundred feet breathing mixed gases, but it takes years of training and is a logistical nightmare. The danger isn't going down — although that certainly is dangerous — it's coming back up. For a scuba diver, a one-hour plunge at two hundred feet breathing regular compressed air would require a ten-hour ascent to purge the deadly levels of nitrogen gas in the blood that accumulate on the way down. A three-hundred-foot ascent on compressed air would most likely kill you.

My best bet in the short term was to talk to William Trubridge. He dives to three hundred feet all the time. Trubridge and other freedivers who use nothing but their bodies to reach this depth have a physical advantage over scuba divers: decompression sickness doesn't affect them. There simply isn't enough nitrogen in a single breath to bubble the blood. At the surface, this nitrogen is quickly purged from the system in a matter of seconds — another function of the Master Switch.

Between 2007 and 2010, Trubridge broke fourteen world records (mostly his own) in the disciplines of constant weight without fins and free immersion. Today he is considered the world's top no-fins freediver, so he knows as much about the experience of diving down to three hundred feet as anyone who's ever lived.

"FREEDIVING IS AS MUCH A mental game as a physical one," says Trubridge. We are sitting poolside at the Messinian Bay Hotel the day after my freediving lesson with Prinsloo. Trubridge, with his cropped hair, wraparound dark glasses, and a worn T-shirt, fits right in with the rest of the freedivers gathered here. He's got the quiet, nerdy energy of a software engineer.

Like almost all competitive divers, Trubridge says he dives with his eyes closed. He'll open them for a moment when he reaches the plate at the bottom of the rope, but that's it. By diving blind, he prevents his brain from using up the energy — and oxygen — it would take to process visual information.

So, Trubridge can't tell me what it *looks* like at three hundred feet down, but he can certainly describe how it *feels*. He leans back in his chair and takes a deep breath. And as he starts talking, my stomach starts tightening once again . . .

In the first thirty or so feet underwater, the lungs, full of air, buoy your body toward the surface, forcing you to paddle as you go down. As you blow air into your middle-ear canals to equalize the pressure, you'll feel a much more intense version of the discomfort you would feel in an airplane as it gains altitude. If you fail to equalize the ears completely, the pressure becomes debilitating, and if you don't return to the surface, you risk damaging your eardrums.

And you've still got 570 feet of swimming to go.

As you descend past 30 feet, you feel the pressure on your body double and your lungs shrink. You suddenly feel weightless, your body suspended in a gravityless state called neutral buoyancy. Then something amazing happens: as you keep diving, the ocean begins pulling you down. You place your arms at your sides in a skydiver pose, relax, and effortlessly dive deeper.

At 100 feet, the pressure triples. The ocean's surface is barely visible, but you're not looking anyway. You've closed your eyes at the surface. Your skin cools as you prepare for the deep water's tightening clutch.

Farther still, at 150 feet, you enter a dream state caused by heightened levels of carbon dioxide and nitrogen in your bloodstream. For a moment, you can forget where you are and why.

At 250 feet, the pressure is so extreme that your lungs shrink to the size of fists and your heart beats at less than half its normal rate to conserve oxygen. Heart rates of freedivers at this depth have been recorded as low as fourteen beats per minute; some freedivers have reported heart rates of seven beats per minute. These reports are the lowest heart rates for conscious humans ever recorded. According to physiologists, a heart rate this low can't support consciousness. And yet, according to the divers, somehow, deep in the ocean, it does.

At 300 feet, the Master Switch really kicks in. The walls of your organs and vessels, working like pressure-release valves, allow the free flow of blood and water into the thoracic cavity. Your chest collapses to about half its original size. During a no-limits dive in 1996, Cuban freediver Francisco Ferreras-Rodriguez's chest shrank from a circumference of fifty inches at the surface down to twenty inches by the time he reached his target depth of 436 feet.

The effects of nitrogen narcosis at 300 feet down are so strong that you forget where you are, what you're doing, and why you're in this dark place, fumbling around. Hallucinations are common. One diver told me that during a very deep dive, she forgot that she was underwater. She began to have strange thoughts about her dog. She pictured herself in a dark park looking for him. As she headed back toward the surface, and the haze of nitrogen narcosis faded, she remembered that she didn't own a dog.

Nitrogen narcosis affects more than just your brain; it affects your entire body. You lose motor control. Everything around you appears to slow down.

Then comes the really hard part. Your dive watch beeps, alerting you that you've reached your target depth at the plate attached to the end of the rope. You open your eyes, force your semiparalyzed hand to grab a ticket from the plate, and then head back up. With the ocean's weight working against you, you tap your meager energy reserves to swim toward the surface. If you lose concentration now, cough, or even slightly hesitate, you could pass out. But you don't hesitate or slow down. You hurry and kick back toward the light.

As you ascend to 200 feet, 150 feet, 100 feet, the Master Switch slowly reverses its effects: the heart rate increases, and the blood that flooded into your thoracic cavity now floods back out into your veins and arteries and organs. Your lungs ache with an almost unbearable desire to breathe; your vision fades; and your chest convulses from the buildup of carbon dioxide. You need to hurry or you'll black out. Above you, the blue haze transforms into a sheen of sunlight. You're going to make it. The air in your lungs is now rapidly expanding, and your body is desperately trying to pull oxygen from the lungs and feed it into your blood. But there is no oxygen to pull; you've already used it all up. Your body literally starts being sucked inward. If this vacuum grows too strong, you will black out. You can stay submerged in a blacked-out state for about two minutes. At the end of two minutes, your body will wake itself up and breathe one last time before you die. If you've been rescued and carried to the surface by the time you take your last gasp, you'll inhale much-needed air and will probably survive. If you are still underwater, your lungs will fill with water and you'll drown. Ninety-five percent of blackouts happen in the last fifteen feet, usually as a result of this vacuum.

But it's not going to happen to you. You've learned well, and you know to exhale most of your air as you get to within about ten feet of the surface.

Some three minutes after you started down, you pop your head up out of the water; the world spins; people yell at you to breathe. You take off your goggles, flick an okay sign, and say, "I'm okay."

And then you get back in the boat and head to your hotel room.

UNTIL 2009, ONLY TEN FREEDIVERS in the world had reached the three-hundred-foot limit in the freediving discipline called constant weight (CWT), which allows the diver to use a monofin — a three-foot-wide wedge of plastic attached to neoprene boots. This Thursday in Greece, the second day of the world championship, fifteen competitors will be attempting that depth.

British diver David King will be one of them. King surprised everyone the night before by announcing that he would try a 102-meter (335-foot) dive, which, if he succeeded, would set a new national record in the United Kingdom. According to his teammates, he hadn't gone deeper than eighty meters in the past twelve months. Progress in freediving is made meter by meter, several freedivers told me yesterday. Attempting to better a record by more than seventy feet is not only audacious but borderline suicidal.

This morning, the waters of Messinian Bay are gray and wind-chopped from a storm that blew through yesterday. It's not raining now, but clouds loom overhead, and subsurface visibility has diminished to about forty feet.

I take a seat at the bow of Georgoulis's boat, next to Prinsloo, who will be coaching her friend Sara Campbell, a women's freediving champion from the United Kingdom who will be making her own world-record attempt a little later. Meanwhile, on the line directly below me, David King takes the last few breaths before his dive. The judge starts the countdown. King dunks his head, upends, and kicks his monofin violently. His silhouette fades into the gray water below. In about ten seconds, he is gone.

The official follows King's descent: "Fifty meters, sixty meters, seventy meters . . ."

"My God, he is *flying* down," says Prinsloo. Speed isn't necessarily a good thing in freediving, she reminds me. The faster King goes, the more energy he burns and the less oxygen he'll have for his ascent.

"Eighty meters, ninety meters . . ." the dive official says. King is now traveling so fast that the official has trouble keeping up. "Touchdown," he announces, and King starts coming back.

"Ninety meters, eighty meters." Then the official pauses. King is coming up at about half the speed of his descent. This is troubling; King will need to ascend faster or he'll run out of oxygen.

"Sixty . . . fifty . . . forty meters." The gaps between the announcements lengthen. Then the official stops altogether. A few seconds later he repeats: "Forty meters." Ten seconds pass in silence. King has now been underwater for more than two minutes.

"Forty meters," the official repeats again. King has stopped, it appears. A sickening anticipation sets in. I look around the sailboat. The officials, divers, and crews all stare at the choppy water and wait.

"Thirty meters."

King appears to be moving, but too slowly. Five seconds pass.

"Thirty meters," the official repeats.

"Oh God," Prinsloo says, holding her hand over her mouth. Five more seconds. The official is staring at the sonar screen, but he's no longer announcing. In the water, we see nothing — no sign of King, no ripples at the surface.

"Thirty meters." Silence. "Thirty meters."

"Blackout!" a safety diver yells. King is unconscious some ten stories below the surface. The divers kick down into the water.

"Safety!" the judge yells. About thirty seconds later, the water around the line explodes in a cauldron of foam. The heads of two safety divers reappear. Between them is King. His face is bright blue, and he's not moving. His neck is stiff.

The divers push King's head out of the water. His cheeks, mouth, and chin are slicked with blood. "Breathe! Breathe!" the divers yell. No response. Bright drops of blood drip from King's chin into the ocean.

"Safety! Safety!" the judge yells. A diver puts his mouth over King's blood-covered mouth and blows. "Safety now!" the judge yells. King's coach, Dave Kent, is yelling into King's ear, "Dave! Dave!"

No response. Ten seconds pass, and still nothing. Someone yells for oxygen. Someone else for CPR. Georgoulis screams, "Why isn't anyone calling a medic? Get a helicopter!" Georgoulis is yelling at me, at Prinsloo, at no one in particular. "What the fuck is going on here?" he shouts.

Behind us, on line one, another diver heads down. Then another surfaces, blacked out. The safety divers move King's supine form to the flotilla and put an oxygen mask to his face. Still no response. His neck is stiff; his facial muscles are frozen into a sickly smile; his eyes are wide and lost, staring into the sun.

*King is dead.* That is the consensus on the sailboat. But we are forty feet away from him now, and through all the yelling, nobody can tell what's really happening. The safety crew on the flotilla are pumping King's chest, tapping his face, yelling at him.

"Dave! Dave?"

Around the flotilla another diver submerges, and another lifts his head to the surface. The competition just keeps going on. I turn to the side of the boat so I can look away. A Czech diver stares at me, closes his eyes, and goes back to mumbling a mantra in preparation for his dive.

Then, miraculously, King's fingers quiver, his lips flutter, and he breathes. Color returns to his face; his eyes open, then softly close again. His limbs loosen. He is breathing deeply, tapping his coach's leg as if to say, *I'm okay, I'm okay*. A motorboat arrives. The safety crew carefully places King in its bow.

As the motorboat carrying King takes off for shore, Trubridge attempts a 387-foot dive on line one, but he turns around early and fails his surface protocol. Next, British contestant Sara Campbell turns back after just seventy-two feet of her world-record attempt. "I couldn't do it," she says, hopping back on the sailboat. She was too shaken by King. There's another blackout on line two. Then another on three.

"My God, this is getting messy," says Campbell. The west winds grow stronger, chopping the ocean, fluttering the sail above us. "It's like dominoes. Everything's falling apart. This is the worst I've ever seen." And yet the competition goes on for three more hours.

On the last dive of the day, a Ukrainian, new to the sport, attempts a beginning descent of forty meters. He dives, surfaces, and removes his mask — and a stream of blood gushes from his nose. He completes the surface protocol and is awarded a white card, meaning the dive is accepted. Blood is okay.

THAT NIGHT AT THE HOTEL, the divers cavort; some laugh, others casually shake their heads at all the drama. Of the day's ninety-three competitors, fifteen of them attempted dives of a hundred meters or more. Of those, two were disqualified, three came up short, and four blacked out — a 60 percent failure rate. King is in the hospital. Nobody knows for sure, but the rumor is that the pressure tore his larynx, which is fairly common on deep dives and, they say, a minor injury.

Regarding the day's events in general, the contestants are less blasé. "This kind of thing never happens," they insist over and over that night in the courtyard, rolling their eyes. It sounds like a practiced response. This kind of thing probably happens all the time, it's just that nobody here wants to admit it. The challenge now is to see who among them can erase today's "messy" events from their minds and dive to even greater depths on the final day of competition.

One person who seems unfazed is Guillaume Néry, a twenty-nine-year-old French freediver and the winner of yesterday's CWT competition. The day after King's near drowning, I meet Guillaume midmorning at a table crowded with other members of the French team.

"I was not there, so don't know exactly," he says in his thick accent. "But I think the main mistake is not for Dave King but for all freedivers. They were focused on this one-hundred-meter number and not on their feelings, not what they really want to do." Néry, who started freediving at fourteen, gained international fame in 2010 with the release of *Free Fall,* a short film that follows him on a thirteen-story freedive in the Bahamas. Since its release, the clip has been viewed on YouTube more than thirteen million times.

"I learned long ago that patience is the key to success in freediving," he says. "You have to forget the target, to enjoy and relax in the water." Néry smiles and runs his fingers through his mop of sandy hair, mentioning that he hasn't blacked out in more than five years of steady freediving. "What is important now is trying to do the dive, surface, and have a smile on my face."

SATURDAY, THE FINAL DAY OF competition, brings scorching sunshine, still air, and clear, calm waters — perfect conditions. The discipline today is free immersion, where divers are allowed to pull themselves down the line to reach their target depth. Free-immersion dives are a little shallower than CWT dives, but they can take a while, sometimes more than four minutes, making them excruciating to watch. The divers got a wrist slap last night from event director Stavros Kastrinakis, who told them, "Dive your limits." The announced dives today appear to be more conservative. Still, there are a number of world- and national-record attempts planned.

"Two minutes," the official from the flotilla announces to the divers. The first competitor is slowly towed by his coach to line three. The diver turns over and descends into the clear water, pulling himself along the guide rope. He touches down and begins the ascent. As usual, the official narrates his progress: "Thirty meters . . . twenty meters."

Another blackout. The safety divers plunge down. Moments later, they bring the diver up. His face is blue; his mouth open. I turn to walk below deck, no longer interested in watching this sport. But seconds later, the diver shakes his head and smiles, then apologizes to his coach.

"See, that wasn't so bad," says Prinsloo, standing behind me on the sailboat. It wasn't, or maybe I'm just getting used to seeing unconscious bodies pulled from sixty feet down. Either way, I return and watch as the next dozen divers all make their dives without incident. Then the elite divers start going; Malina Mateusz of Poland breaks a men's national record with a dive of 106 meters. The women's reigning world champion, Russian Natalia Molchanova, sets a world record of 88 meters. Antoni Koderman dives 105 meters to set a new Slovenian record. Néry breaks the French record with 103. Trubridge does 112, almost effortlessly. Seven national records are broken in an hour. Everyone is in control. The sport, again, is beautiful.

Then, at line two, a commotion breaks out. The safety divers have lost a Czech contestant named Michal Rišian. Literally lost him. He's at least two hundred feet down, but the sonar is no longer picking him up. He has somehow drifted away from the rope.

"Safety! Safety!" yells the judge. The safety divers go down but come up a minute later with nothing. "Safety! Safety! Now!" Thirty seconds pass. There is no sign of Rišian anywhere.

On line one, Sara Campbell is preparing to dive. From below her, three and a half minutes after he went down on line two, Rišian emerges — some forty feet away from the line he was first attached to.

There's confusion. Campbell jerks away, frightened. Rišian snaps off his goggles, saying, "Don't touch me. I'm okay." Then he swims back to the sailboat under his own steam. He plops down beside me on the hull, laughs, and says, "Wow, that was a weird dive."

That's one way of putting it. Before Rišian's dive and per the usual routine, his coach attached the lanyard around Rišian's right ankle to the line. As Rišian turned and plummeted, the Velcro securing the lanyard came loose and fell off. The safety divers saw it floating, unattached, and rushed down to stop Rišian, but he was already gone, a hundred feet deep. Rišian, unaware, closed his eyes, meditated, and drifted downward. But he wasn't going straight down — he was angling 45 degrees away from the line, into open ocean.

Rišian's coach, realizing that death was the likely outcome of this screwup, floated motionless at the surface, gazing at the safety divers, who were too stunned to blink. "I'll remember their looks for a long time," he said later. "Terror, awe, fear, and sadness." Meanwhile, Rišian was diving farther down and farther away, oblivious to his peril. At 272 feet, the alarm of his dive watch sounded. He opened his eyes and reached out to grab the metal plate, but there was no plate. "I couldn't see any tickets, any plate, any rope, nothing," he said. "I was completely lost. Even when I turned up and looked around, I saw only blue."

When you're twenty-seven stories down, in cloudy water, all directions look the same. And they all feel the same — the water pressure makes it impossible to gauge whether you're swimming up, down, or sideways.

For a moment, Rišian panicked. Then he calmed himself, knowing that panic would only kill him faster. "In one direction there was a bit more light," he told me. "I figured that this is where the surface was." He figured wrong. Rišian was swimming horizontally. But as he swam, trying to remain conscious and calm, he saw a white rope. "I knew if I could find the rope, I would be okay," he said.

The chances of Rišian finding a line two hundred and fifty feet down – especially one so far from his original line of descent – were, I would estimate, about the same as hitting 00 on a roulette wheel. Twice. But there it was, the line Sara Campbell was about to descend, some forty feet away from where he had first gone down. Rišian grabbed it, aimed for the surface, and somehow made it up before he drowned.

ON THE FINAL NIGHT, the divers, coaches, and judges gather on the beach for closing ceremonies. Strobes and spotlights glare from an enormous stage, Euro pop blasts from a DJ booth, and a crowd of a few hundred dance and drink beneath a night sky sequined with stars. Behind the stage a bonfire rages, heating the bare, wet bodies of those who couldn't resist one last splash.

The winners are announced. All told, the divers broke two world and forty-eight national records. Competitors also suffered nineteen blackouts. Trubridge won gold in both constant weight and free immersion.

"Rišian is the real winner here," says Trubridge, sipping a beer beside his wife, Brittany. Behind us, every twenty minutes or so, a video screen shows the chilling footage of Rišian's tetherless dive, which was recorded by underwater cameras. At the end of the video, the crowd cheers, and Rišian, who's now drunk on "birthday" drinks (to celebrate his new life after his near-fatal dive), rushes to the stage to take a bow. Dave King, the diver who suffered the horrific blackout just two days ago, walks through the crowd with the British team, smiling and seemingly in perfect health. Néry, in quintessential French style, is smoking a cigarette.

"There is such a strong community here," says Hanli Prinsloo, drinking a cocktail by the bonfire. "It's like all of us, we have no choice. We have to be in the water; we've chosen to live our lives in it, and by doing that, we accept its risks." She takes a sip.

"But we also reap its rewards."

# −650

A MONTH LATER, I'M INVITED to view a different kind of freediving, one with a purpose. A handful of freelance researchers plan to spend ten days diving, studying, and attaching tracking transmitters to the dorsal fins of man-eating sharks. It's all taking place along the coastal waters of an island on the other side of the globe that I've never even heard of. Getting there is the first challenge.

IT TAKES FIFTEEN HOURS, THREE meals, four small bottles of wine, seven films, and five trips to the bathroom to fly from San Francisco to Sydney, Australia. Next, there is a four-hour layover in Sydney International Airport (one bagel, a twenty-minute nap on the floor, one bag of cashews, forty-five minutes at the newsstand reading *Rolling Stone*) before the connecting flight to Saint-Denis, the capital of Réunion. The airplane is an old Airbus A330, a model infamous for inflight malfunction, and the paint job makes it look like it's from the 1980s. The interior is just as shabby: seats are stained; the handles of the overhead luggage compartments are loose and scratched, and their color, once white, has faded to yellow. The cabin is only 20 percent full, occupied mostly by

elderly couples. Everyone except me speaks French. Within an
hour after we take off, passengers are splayed across empty rows,
fast asleep. More wine, more movies, more meals. Night dissolves
into day.

Twelve hours later, the seat-belt lights come on. The airplane
arcs west, and looking through the portside windows, I see a tiny
island appear in the distance. The captain noses the plane down,
and an otherworldly landscape appears: Crowns of miles-high
volcanic peaks poke through billowing white clouds. Blue water
laps on white beaches. Forty-story waterfalls spray mist on green
jungle floors. It's such a clichéd scene of tropical wonder that it
might as well be computer-generated, like a backdrop from *Ju-
rassic Park: The Lost World*. But this isn't a movie set or a screen-
saver. The exotic, prehistoric landscape below is what France
looks like four thousand miles from Paris.

Réunion is the southernmost outpost of the French Republic and an outermost region of the European Union. At just 970
square miles, about a quarter of the size of the big island of Hawaii, it's a tiny dot situated six thousand miles west of Australia
and about four hundred miles from the east coast of Madagascar.
The French came here in the 1600s, named it Bourbon Island,
and used it as a trading post and sugar plantation for the next
few centuries. Today, Réunion is to the French what Hawaii is to
Americans — a tropical getaway with all the modern conveniences
of the mainland but none of the chilly weather. They come here
for the same reasons: to retire, start a new life, honeymoon, or
thaw out during long winters. Réunion's biggest claim to fame
is that, in 1966, it got seventy-one inches of rainfall in a single
twenty-four-hour period — a world record. In 1671, the population on the entire island was ninety. In 2008, it was more than
eight hundred thousand. Although the French claim Réunion as
their own, immigrants from neighboring India, China, and Africa
now predominate. Most residents live along the west coast near
a string of former colonial outposts. There's a Catholic church in
every town center, and a beach bordered by grids of colorful low-
rise houses. The local beer, called Dodo after the extinct bird that

was (wrongly) thought to have inhabited Réunion, tastes vaguely of soap.

The food, however, is delicious—Parisian quality with an African twist. The weather is always warm and inviting. The landscape and beaches are uncrowded, pristine, and as spectacular as any South Pacific island's. Réunion would be a paradise if it weren't for one notable problem: the constant threat of being eaten by sharks.

In recent years, for reasons nobody can explain, shark attacks have been on the rise. In 2010, bull sharks suddenly started going on a rampage, killing and mauling swimmers and surfers along the island's most luxurious beaches and resorts.

The worldwide average for shark fatalities is six a year; two deaths and a half a dozen injuries occurred on tiny Réunion in the span of three months. It was the most dramatic increase in shark attacks that Réunion had ever seen, and it threatened to destroy the island's delicate tourist-based economy.

It was particularly vexing for Fred Buyle, a photographer and shark conservationist I'd met at the freediving world championships in Greece. He called me a week after I returned home to discuss the upside of freediving—the side without bleeding mouths or death throes. He explained how useful it can be in shark research.

"Freediving, it is also a tool," he told me over the crackling phone line in his lilting French-Belgian accent. It was a way to get in touch with the ocean's animals and, he hoped, to help save them.

I first met Buyle in a hotel bar in Kalamata, with a small group of other freedivers. When I asked what he did for a living, he demurred. "I freedive some," he said, "and I take a few pictures." It wasn't until I Googled him later that night that I discovered he was legendary—one of the earliest competitive freedivers and one of the most sought-after underwater photographers in the world. Websites were filled with photographs of him diving inches away from great white sharks, swimming among swirling schools of hammerheads, and arm to fin with whitetips.

Buyle was traveling to Réunion, he said, to stop a shark slaughter. In the wake of the most recent attacks there, angry locals were trying to catch and kill the entire local bull shark population. This would decimate Réunion's pristine ocean ecosystem.

His plan was to join a troop of voluntary marine researchers, including a Réunion-based engineer named Fabrice Schnöller, and freedive down about eighty feet, to the seafloor. There he would place satellite tags on the bull sharks' dorsal fins. These tags would track the sharks' swimming patterns and locations, alerting locals if they came too near shore. It would be the world's first real-time shark-tracking system.

Buyle believed that the recent wave of attacks were accidents. Bull sharks, he said, don't like to eat people. There had to be some other reason they were approaching shore. By tracking their movements, his team might be able to identify the cause, help remedy it, and save the bull sharks from annihilation.

One could understand why locals might see things differently. Bull sharks are among the toughest and deadliest marine predators. They can grow up to twelve feet long and weigh as much as five hundred pounds. Their highly evolved kidneys allow them to flourish in both fresh and salt water, and they have been observed in a variety of extreme environments: over two thousand miles up the Amazon in the foothills of the Peruvian Andes, in floodwaters along city streets in eastern Australia, and on seafloors at six hundred and fifty feet deep. They eat almost anything with a face — fish, other bull sharks, sea turtles, birds, dolphins, crabs. Along with tiger and white sharks, bull sharks are responsible for more attacks on people than any other shark species on Earth.

Like most sharks, bulls spend much of their time in deep water, where visibility is extremely limited or nonexistent. This makes them almost impossible to study. Submarines, robots, and divers in atmospheric diving suits can make it to six hundred and fifty feet and deeper, but those devices don't allow enough speed or flexibility for a diver to follow the bull sharks, tag them, or make any worthwhile observations. Even on the sunniest days, in the clearest water, the ocean at that depth — called the twilight or mesopelagic zone, which extends from 650 feet down to 3,300 feet—gets less than 1 percent of the light at the surface. That's not enough to support photosynthesis, and as a result, food at these depths is scarce.

Bull sharks adapt by hunting for prey in shallower water, then returning to deeper waters to migrate. The only way to conduct long-term studies of them—and almost any other kind of shark—is to wait for them to feed closer to the surface and tag them with tracking devices that will monitor their movements on the way down.

Tagging isn't easy, however. Scuba diving or tagging from a boat is dangerous, and sometimes it just doesn't work. The sharks get nervous and swim away, or they're injured in the process. Sometimes they bite.

Buyle told me the safest, most effective way to tag Réunion's sharks is to meet them on their own terms by freediving down deep enough to slap on a transmitter. Still, he acknowledged, it was a risky operation with no guaranteed results.

He would meet me in Réunion in three weeks.

A SHORT HISTORY OF THE mesopelagic.

In 1841, a British naturalist named Edward Forbes dredged samples from the deep water along the Mediterranean and Aegean Seas and came up empty. Not a shell, plant, fish, or any sign of life. Forbes declared the water below nine hundred feet to be a black desert, and he named these depths the azoic ("lifeless") zone. His declarations held, uncontested, for two decades.

Then, in the 1860s, a contrarian scientist in Norway decided to check Forbes's work. Michael Sars sailed out to the middle of the Norwegian Sea, dropped some nets and buckets down a few hundred feet, and then hoisted them back up. He did this again and again, and he discovered that the "lifeless" depths were full of life. Within a few years, Sars had found more than four hundred animal species in this netherworld, some of them discovered as deep down as twenty-five hundred feet. The most startling discovery was the sea lily—a flower-shaped animal with a long stalk of a body and a crown of petal-like pinnules that, scientists believed, had thrived in the time of the dinosaurs, one hundred million years ago.

The sea lily had long been thought to be extinct, but there it was, in a wooden bucket on the deck of Sars's boat, clearly flourishing a thousand feet down. The deep sea, Sars posited, was not only abundant with life but also a link to our planet's ancient past. And the farther down we went, the farther back in time we reached. While land life was in a state of constant tumult, ravaged by storms, earthquakes, floods, droughts, meteors, and ice ages, nothing much seemed to roil the deep water. Every day featured the same dim blue light; every night was inky black. The weather never changed. It was a living museum.

A decade after Sars's discoveries, scientists had found more than 4,700 new species in the deep sea. They had also sounded the seafloor and mapped out a geography as dramatic as any on land — wide-open plains, rolling mountains, and valleys five miles deep.

A more accurate view of the ocean's depths was emerging, but it was still rudimentary at best — the equivalent of exploring life on land by lowering a butterfly net over the side of a hot-air balloon at night. The mesopelagic, or "middle" zone, now had a proper name, but it still had no face. Nobody had actually *seen* what it looked like, and nobody knew what really went on down there.

It took another thirty years for someone to take a picture. The man who did it was William Beebe, a researcher at the New York Zoological Society. Beebe had no engineering experience and had never seen a vessel capable of plummeting hundreds of feet down into the ocean, but that didn't deter him. He designed a deep-sea machine called a bathysphere (Greek for "deep sphere") and parked it just off the coast of Nonsuch Island, Bermuda. In June 1930, he prepared it for its first manned submersion.

The bathysphere was essentially a large, hollow cannonball with three three-inch-thick fused-quartz windows and a four-hundred-pound entrance hatch on top. It was just big enough to hold two men with one kneeling on his heels and the other sitting directly in front with his legs drawn up. A steel cable, attached to the roof and wound around a mechanical winch, was used to lower it into the water and bring it up again, like a yo-yo. Canisters of compressed air supplied oxygen; air conditioning came in the form of palm-frond leaves that would be used as a fan.

Things went wrong all the time. On unmanned test dives, the roof cables would tangle and bind. In strong currents, it would swing and sway violently, flinging objects around the cabin. Sometimes it leaked.

Once, Beebe and the boat's crew hoisted the vessel on deck after a test run and saw through one of the windows that it was completely filled with water. As Beebe began loosening the top hatch, a bolt shot across the deck, leaving a half-inch indentation in a piece of steel thirty feet away. From the bolt hole, a stream of water shot out with such force that, in Beebe's words, "it looked like hot steam." He realized that the bathysphere must have taken on water at extreme depths, and as it was hoisted back to the surface, the pressure inside it had steadily mounted, reaching upwards of thirteen hundred pounds per square inch. The loosened bolt shot out like a bullet. If Beebe had been inside the bathysphere during the dive, he would have drowned.

Dangers be damned. On June 6, 1930, he crawled into the bathysphere and readied himself for the first dive. Beside him was a Harvard engineer, Otis Barton, who had done much of the design work for the vessel and raised most of the money to build it. The crew released the winch, and the bathysphere splashed into the water. The cable spooled out, and Beebe and Barton disappeared.

By the time the men had descended three hundred feet, the cabin had sprung a leak. Beebe decided to continue on. At six hundred feet, a shower of sparks erupted from a light socket. Barton pressed on the wire fitting, and the sparks stopped. The bathysphere sank deeper.

Beebe and Barton watched as the water darkened around them, like houselights slowly dimming before a performance. "I pressed my face against the glass and looked upward and in the slight segment which I could manage I saw a faint paling of the blue," Beebe would later write. "I peered down and again I felt the old longing to go further, although it looked like the black pit-mouth of hell itself."

The depths were alive with fantastic creatures — fish, gelati-nous orbs, and never-before-seen life forms. As they approached seven hundred feet, the water wasn't black, as Beebe had thought it would be, but a dusty blue. "On earth at night in moonlight I can always imagine the yellow of sunshine, the scarlet of invisible blossoms," he wrote. "But here, when the searchlight was off, yel-low and orange and red were unthinkable. The blue which filled all space admitted no thought of other colors."

On their initial manned dive, Beebe and Barton became the first men to lay eyes on the deep, blue world of the mesopelagic, diving down just over eight hundred feet.

And yet, being inside the bathysphere was a hopelessly iso-lating experience. Beebe and Barton could catch glimpses of the animals of the deep, but, dangling from a steel cable, they were unable to follow, interact, or study them in any significant way. They could barely even take photographs. They proved that ani-mals actually existed at great depths — Beebe and Barton would eventually make it to 3,028 feet, over a half a mile down — but be-yond that, they knew little of where this alien life went, what it ate, or how it could navigate through the dark, featureless waters of the deep ocean.

That began to change in the 1940s and 1950s, when research-ers started tracking marine animals with plastic identification tags. While it was impossible to track or study animals that stayed down in the mesopelagic, vertical feeders, like sharks, that spent much of their time in the mesopelagic but often came to the sur-face to feed, could be tagged.

A shark tagged in one area and observed in another would show scientists how far sharks were migrating and where they were going. Some researchers captured sharks, slit their abdo-mens, inserted tags, sewed them up, and released them back into the water. These tags could last for decades. (One inserted into a shark in 1949 was discovered forty-two years later.) In a U.S. campaign started in the late 1950s and continuing for about thirty years, some 106,000 sharks in the northwest Atlantic were fitted with tags.

By the 1960s, researchers started tagging sharks with transmitters, which, for the first time, offered immediate data about how fast they swam, where they went, how far, and at what depth.

The results were startling. Half of all known species spent much of their time in the cold and dark waters of the deep ocean. At that depth, they'd migrate thousands of miles in schools of hundreds, swimming head to tail in perfect unison, following an invisible line. Then they would return to their point of origin, following the same invisible line with the same precision.

Even in the clearest tropical ocean, there is very little light at 650 feet down, and there was nothing to feel or smell or see that could help the sharks find their way. And yet they seemed to know where they were and where they were going at all times. For a human, that would be like putting on a blindfold and earplugs and walking three thousand miles from Venice Beach, California, to Coney Island and then back again. And doing it every year.

AROUND THE SAME TIME MARINE researchers were scratching their heads over these new findings, a German zoologist named Friedrich Merkel heard about some peculiar behavior among European robins. Merkel's colleagues had witnessed the robins hopping in the same direction that they naturally migrated in. The birds continued this directional hopping even in enclosed areas, where they couldn't take cues from the sun or sky. It was as if the birds had an innate sense of their location and destination, even when they couldn't see anything.

In 1958, Merkel gathered a flock of robins and placed them, one at a time, inside a chamber about the size of a wash bucket that blocked the sky, stars, and sun. The floor of the chamber was covered with a touch-sensitive electric pad that recorded the direction the robin hopped in. Over several months, Merkel observed their movements. The results were always the same: In the spring, the robins hopped north; in the fall, they'd hop south. In other words, the robins hopped in the exact directions of their normal migration routes.

Merkel repeated the tests in various chambers under various conditions and got nearly identical results – with one exception. When he put the robins in a magnetically shielded chamber, their sense of direction disappeared.

On a compass, the pull toward magnetic north is a reaction to the Earth's magnetic field – positive and negative charges created by the circulating molten iron in the planet's core. For Merkel and his colleagues, these experiments provided ample proof that robins had a magnetic sense of direction. Other scientists balked, claiming the data was weak. The idea that birds, animals, or *any* creatures could orient themselves by the subtle energy of magnetic fields – using a sense other than vision, hearing, feel, taste, or smell – was just too weird a proposition for most scientists to accept.

But Merkel was right.

Twenty-five years after his experiments, this magnetic sense (which became known as magnetoreception) was shown to exist in bacteria, and shortly after that, scientists found overwhelming evidence that other creatures used it as well, including birds, bees, ants, fish, and sharks.

Experiments for magnetoreception in humans conducted over the next thirty years suggested that we too might have this sixth sense. But to prove it, scientists needed to know exactly how it worked in the human body. To do that, they needed a sensory receptor. They'd find a likely candidate in 2012.

FRED BUYLE WALKS THROUGH THE security gates of Roland Garros International Airport in Saint-Denis, the capital of Réunion, pushing a cart full of spear guns and dive equipment. Above him, flocks of bats and small black birds lost in the rafters fly lazy figure eights. The ammoniac smell of bird and bat excrement mixes with the sticky, steamy tropical air.

A throng of reporters waits at the exit gate, cameras rolling. In the past few days, local media have painted Buyle as a kind of shark whisperer. Wearing a tight black T-shirt and sporting the shaved head and musclebound physique of Mr. Clean, Buyle is visibly annoyed by the reporters' presence. He politely exchanges a few quick words with them in French, then pushes his way through the exit doors to Fabrice Schnöller's silver pickup truck. "This is bullshit," Buyle says in his resonant monotone. He hops into the passenger seat. "There's no hero here. No quick solution. It's just the beginning of a long process."

That evening, Buyle, Schnöller, and I take my bite-size rental car through Saint-Denis's maze of narrow cobblestone streets and soot-covered colonial buildings. Before long, we arrive at a restaurant overlooking a beach with a perfect curling wave. It's an eerie scene: a glassy, head-high wave on a tropical island at sunset with nobody surfing it. In fact, there's nobody on the beach at all.

"It's illegal to be on the beach now. You swim in the water, they will put you in jail," says Schnöller, taking a seat at the patio table. Schnöller used to own a lumber store down the road, but he sold it five years ago, after having a spiritual experience diving with sperm whales. He now spends his time running Dare-Win (short for Database Regional for Whales and Dolphins), a nonprofit organization focused on dolphin- and whale-communication research. With his uncombed swatch of short gray hair, oversize multicolored shorts, and wild gesticulations, he is the dervish to Buyle's monk.

Schnöller orders a beer and leans back in his chair. He mentions that travel agencies are now warning travelers away from Réunion — it's too dangerous. "Nobody wants to be responsible, and the government bans [people from] the beaches to avoid paying costs of amputation, rehabilitation, whatever." He sighs. "I mean, even the locals are scared of these sharks."

In September 2011, a surfer had his leg bitten off by a shark.

A week later, a shark charged a kayak, hitting it from beneath the prow and sinking it. The kayaker was picked up by a passing boat and survived. Then a thirty-two-year-old former body-boarding champion was dragged from his board in a crowded line and half devoured in less than thirty seconds. His mutilated body washed up on the shore. Two months after that, a spearfisher wading in chest-high water was bitten on the ass.

"It's very crazy," says Buyle. "Sharks don't like to eat humans. That's what was so weird. Something was frightening them maybe, bringing them close to shore. But what?"

Schnöller takes out a pen and begins drawing on the back of a napkin. "This is how we will find out," he explains, pointing at a boxy figure with some circles around it. It's a picture of the shark-tracking system he's invented, which he calls SharkFriendly. Six months earlier, when Schnöller first met Buyle at an underwater film festival in Paris, the two discussed working together on a shark-tagging project on Réunion. After the recent spate of shark attacks, Schnöller developed his first schematic for Shark-Friendly. The two have been working out the details ever since.

SharkFriendly is an acoustic system that follows a shark in real time. Most tagging systems work with satellite technology — a tiny computer built into a metal tube the size of a cigar, which adheres to the shark for six to nine months, then detaches, floats to the surface, and uploads collected data to a satellite. While accurate, satellite tags offer only backstory: what the shark did last year, last month, last week, but not what it is doing right now. "They provide incredible information, but it's all history," Schnöller says of existing systems. He believes Réunion's surfers and swimmers need to know where the homicidal sharks are now, not where they were yesterday.

SharkFriendly relies on a weave of acoustic systems, beacons, and satellites. On the napkin, Schnöller diagrams the particulars of his system, starting with a sketch of the coastline of Boucan Canot, the locus of the recent attacks. When a tagged shark gets about fifteen hundred feet from shore, beacons that Schnöller placed off the coast will recognize the tag's high-frequency signal and relay an alert to a satellite, which in turn will signal a computer server that updates a website and mobile app to warn people that a shark is nearby.

Schnöller tells me that nobody has tried to set up a system like this before. And nobody is paying him and Buyle to do it now. Schnöller crumples the napkin and throws it onto his plate. "But what else are we supposed to do?" he says. "Sit here and do nothing?"

THREE DAYS LATER, SCHNÖLLER and I arrive at La Possession Marina for our third shark-tagging attempt. The past two days at sea were failures. Buyle dove for hours but never saw a single bull shark. Today, we'll try again, in a marine sanctuary close to Boucan Canot, the beach where the body-boarder was eaten just two months earlier. Diving in this area is illegal, but Schnöller and Buyle will take the risk of arrest or injury in order to increase the chances of finding a shark. They've also brought in some backup.

Waiting on the dock beside our motorboat is Markus Fix, a forty-four-year-old German computer programmer and the technical wizard behind SharkFriendly. Fix, who is wearing a T-shirt that says *Science: It Works, Bitches,* has created an underwater system that will broadcast the noise made by an injured fish. Sharks are opportunists, Schnöller tells me, and will never pass up an easy meal. Nothing sounds better to them than injured prey.

Joining Fix is Guy Gazzo, a slender man with salt-and-pepper hair and anchorman good looks. Gazzo is one of the best freedivers in Réunion and can hold his breath underwater for more than five minutes. I'm shocked when Schnöller tells me he's seventy-four. He looks twenty years younger. I say hello. Gazzo replies with *bonjour.* I learn later that Gazzo refuses to speak English because he's still mad at the British for bombing the French navy in Toulon in 1942, when he was five.

Next to Gazzo is William Winram, a Canadian freediver and Buyle's longtime friend. Last year, Winram — who is six three and has a gargantuan frame that makes him look even bigger — set a national freediving record by pulling himself down a rope thirty-two stories deep. I shake his big hand, which feels like I'm gripping a bunch of hot dogs, and hop in the boat.

We leave the harbor and head toward La Possession, a bull shark hotspot. With its rows of houses, beautiful sand beaches, and low-hanging trees, La Possession is made even more scenic by a range of huge mountains that rise a few miles inland. Known as the Cirques, these mountains climb ten thousand feet in a distance of less than ten miles and are so out of proportion with the rest of the local geography that they look like a poorly balanced landscape painting.

About a mile off the coast, Schnöller cuts the motor, and Buyle and Gazzo slip on neoprene gloves, boots, and two-piece wetsuits. They grab their goggles and spear guns and then descend into the diamond-clear water. I watch at the surface as they disappear for several minutes at a time, then return with fish wriggling on the ends of their spears. Winram sits on the back deck, squinting in the bright white morning sun, putting on his wetsuit more slowly. I ask if he's going to join Buyle and Gazzo.

"Yeah," he says. "But I need to take a shit first."

He gets in the water, takes a few big gulps of air, and kicks down eighty feet to the seafloor, where he slips off his wetsuit pants, does his business, and kicks back to the surface. Because of the reverse pull past forty feet, Winram's business will stay stuck to the seafloor instead of rising to float.

Meanwhile, Buyle and Gazzo have returned and are sitting on the deck, slitting the heads of foot-long tarpon they've just speared. They spill the innards into a makeshift strainer that Schnöller constructed from a discarded washing-machine drum he found on the roadside. The strainer will broadcast the smell of fish blood to sharks hundreds of feet away.

While the divers are busy baiting, Schnöller and Fix set up the underwater sound system. Schnöller tells me that sharks have keen hearing and, in the right currents, can home in on prey from eight hundred feet.

"The recording is from 1966 — it's the only one I could find!" says Schnöller, pressing Play on a car stereo that Fix jerry-rigged inside a plastic box. The scream of a crippled kingfish blasts from the speaker, which sounds like someone crinkling a plastic water bottle. Schnöller says he knows an Australian who's proved that sharks are attracted to the music of AC/DC, "You Shook Me All Night Long" in particular.

"What they're listening for is random bursts of low frequencies," he explains. "There's a lot of that in AC/DC." To that end, a little later, Schnöller and Fix will try their own test by blasting the water with tunes recorded by Rammstein, a hard-core metal band from Germany. "The longhaired shark will like it," Schnöller quips.

With the water bloody and thumping with kingfish screams, Buyle fixes an acoustic tag to his spear gun and prepares himself to go deep.

"Come on, James, the water's fine," he calls to me from below. It's 9:00 a.m. and already scorching hot on the boat. A quick dip sounds great. I've been on Réunion for five days now and haven't touched the ocean. I slip into my swim trunks and try not to splash as I get in.

Through my goggles, I watch Gazzo in the distance, descending slowly amid plumes of fish blood into the darker depths, spear gun in hand. Buyle follows, paddling quickly. When he reaches neutral buoyancy, he puts his arms at his side and glides effortlessly down. No matter how many times I see this, it's always an awe-inspiring and spooky thing to watch.

Beside me on the surface, Winram is flailing his arms and legs in the water as though he can't swim, keeping a watchful eye through his mask on the seafloor below. It takes me a minute to realize that he's trying to attract sharks by swimming slowly in a circle, flapping his arms and legs, and acting like a wounded seal. I realize that I've been doing exactly the same thing for the last few minutes. I suddenly feel like I'm standing at an ATM in a bad part of town. I very quietly scramble to the motorboat, pull myself onto the deck, and take a seat in the shade of the canopy, back where I belong.

"*Oui*. Shark!" Buyle says a few minutes later as he surfaces. He calls out to Gazzo and Winram, telling them to dive down in hot pursuit. Fix turns up the volume on the car stereo. Schnöller and I peer over the side of the boat but can't see anything. The divers are too far away. A minute passes. The ocean's surface remains still and flat. Finally, Buyle bursts up, takes a breath, and then kicks back down. I haven't seen Gazzo or Winram for a while, perhaps two minutes. I ask Schnöller what's going on, but he just shrugs and shakes his head.

Eventually all the divers return. Buyle pulls his spear out of the water; the acoustic tag is still stuck to the end of it. Back on the boat, Buyle explains that the sharks were disturbed by all the commotion and left. He, Gazzo, and Winram spend four more hours diving without seeing another shark. At around three in the afternoon, Schnöller starts the motor and we beeline back to the marina.

"They are just so nervous," says Buyle, yelling over the engine as we thump across open ocean toward the port. "It's very unusual," he says. "In Fiji, Mexico, the Philippines, you dive and there are sharks everywhere. You can't help but be around them. But these sharks are different." He exhales. "This could be a challenge."

THE FOLLOWING DAY, AFTER YET another failed mission—sharks spotted, none tagged—I'm knocking at the front door of Buyle's rented apartment, a shabby concrete-block building a half a mile from Boucan Canot. He answers in a T-shirt, bare feet, and shorts, and leads me to a small desk cluttered with cameras, cords, and computers. His laptop screen shows a gallery of photographs of him swimming with hammerheads, whitetips, and other shark species.

"Drop me in the water with some sharks, and I'm happy," he says with a laugh. Buyle starts a video on his laptop. It shows a freediver floating through the gray haze of deep water, slowly approaching a shark the size of a station wagon. The diver, of course, is Buyle, and the shark is a fifteen-foot, four-thousand-pound great white. The video turns my stomach. I tell Buyle it seems like he's asking for trouble.

"Do I look like some adrenaline junkie?" he says, sipping water from a steel bottle, wearing his most monk-like expression. "Skydiving, jumping with bikes. I hate all that shit!" he says. "Freediving with sharks is the opposite of an adrenaline sport. You need to be calm, balanced. You need to know yourself. Being relaxed and in control is the only way you can do it."

Buyle grew up in a small house his father built, just paces from the forty miles of sand and wind-ripped grass that make up Belgium's tiny coastline. His great-grandfather was the official photographer for the king of Belgium in the 1920s. His father was a successful fashion and advertising photographer until, at around age forty-five, he dropped out of the business and toured Europe in a Volkswagen, then married a woman half his age (Buyle's mother) and turned to building sailboats in the garden at the back of the house. Buyle spent his youth playing in those boats and sailing with his father in the gray waters of the North Sea. The family traveled often, usually to exotic tropical locations. Buyle was snorkeling by age seven, spearfishing by ten, and swimming with sharks at thirteen.

"I saw no signs of aggression," he recalls. "I was happy to dive with them." When he was fourteen, he and his friends started freediving. He admits that he knew nothing about it and had no idea how to train. "We had to just find out everything ourselves," he says. "It was an adventure."

In 1988, the freediving film *The Big Blue,* a fictionalized account of the rivalry between freedivers Jacques Mayol and Enzo Maiorca, was released, and freediving's popularity in Europe soared. Buyle, who was sixteen at the time, saw the film as a confirmation. "To me, the film just looked like a documentary of what we were already doing!" he says.

It took Buyle four years of practice to dive down to one hundred feet, a significant depth at the time. After that, he says, everything opened up. By his early twenties, he was diving competitively, and by twenty-eight, he had claimed four world records in the sport, at one point doing a weight-aided dive to 338 feet.

In 2003, during a training session for a world-record weighted dive to more than five hundred feet, Buyle suffered a horrendous accident. He made it down just fine, but as he was about to begin his ascent, the balloon designed to lift him back up didn't inflate properly. He blacked out at two hundred feet. The balloon eventually dragged his comatose body to the surface. He suffered extreme trauma to his lungs but fully recovered after a month and continued doing deep dives.

"To me, freediving was always about exploring the ocean, being a part of it," he says. "It was a way to get to another level, go deeper into the water, push new boundaries." Increasingly frustrated with the competitive, egocentric drive of his fellow freedivers, Buyle quit doing that kind of diving in 2004. "The exploration component was gone," he says. "Freediving became just another sport."

Buyle now spends about two hundred and fifty days a year diving in oceans around the world, filming documentaries, photographing marine animals, lecturing at events, leading freediving tours, and, the thing he loves best, educating the public about sharks. "The fact is, for so long nobody knew anything about sharks," he says. "And humans fear what they don't know." Tagging, Buyle says, can help us allay what he calls our irrational fear of this animal.

His first job tagging was on the island of Malpelo, off the west coast of Colombia, in 2005. Colombian researchers had speculated that hammerhead sharks in the area were migrating as far south as the Galápagos Islands, some fourteen hundred miles away. If they were, Colombia could establish the whole region as a marine reserve and protect the shark, but first the scientists

needed to prove it. They called in Buyle. During three trips span-
ning three years, he dove to depths of more than two hundred feet
and tagged a hundred and fifty hammerheads with both acoustic
and satellite tags. From the data, researchers found that hammerheads were not only migrating to waters around the Galápagos
and farther away but doing so in perfectly organized packs, several hundred strong, in very deep water. The data on smalltooth
sand tiger sharks, a rare species, showed that they were diving
a staggering six thousand feet down and migrating hundreds of
miles and back again. Nobody had any idea that sharks could do
this kind of thing, because nobody had bothered to look. "We
were the first," says Buyle, shooting me a smile. As a result of
these and other conservation efforts, in 2006, 3,300 square miles
around Malpelo were designated as a UNESCO World Heritage
site.

WHILE NOBODY KNOWS EXACTLY HOW hammerheads, feroxes,
and other sharks can navigate in permanently black, deep waters,
most marine researchers believe that tiny bumps on the sharks'
heads and the sixth sense of magnetoreception have something
to do with it. Called ampullae of Lorenzini, after the Italian anat-
omist who described them in 1678, these little bumps are actually pores filled with electrically conductive jelly. At the bottom
of each of the roughly fifteen hundred pores is a hair cell that
resembles one of the tiny hairs inside a human ear. These cells,
called cilia, can pick up the slightest change in electrical fields in
the water. They work in coordination with the lateral line, a series of sensory cells that run down the middle of the shark's back
from nose to tail.

All animals, including humans, generate weak electrical fields
from neurons constantly firing off electrical signals. A shark's
body works like an enormous antenna, tuning in to the signals
pulsing around it. When the shark picks up a signal it likes, it
moves in closer. If the signal seems like something it could eat, it
takes a bite.

Buyle tells me that the full wetsuit he and the other freedivers are wearing isn't just meant to keep them warm — the water temperature in Réunion is a balmy 78 degrees — but also to dampen the electrical signals their bodies send out.*

Sharks' electroreceptive senses are remarkably acute. Tests on captive great white sharks have shown that they can sense electrical fields as small as 125 millionths of a volt. Smooth dogfish sharks can detect 2 billionths of a volt, while newborn bonnethead sharks can detect fields less than 1 billionth of a volt.

To put this in perspective, imagine dropping a 1.5-volt battery in the Hudson River in Manhattan and then running a wire from that battery to Portland, Maine, some three hundred and fifty miles away. The dogfish and bonnethead sharks could detect the faint electrical field coming off the wire. This sense is *five million times* stronger than anything humans can feel. It's by far the most acute sense yet discovered on the planet.

(If these facts are supposed to allay people's fears about swimming with sharks, Buyle and his team may have to work harder on their messaging. Knowing that sharks can track the weakest electrical signals pulsing from my head and heart only makes me fear them more.)

Because sharks' electroreceptive sense is so keen, many scientists believe they can sense the subtle energy of the Earth's magnetic field, which puts out a force of about one-quarter to one-half of 1 percent of the force of a standard refrigerator magnet — significantly stronger than the electrical fields sharks already sense in their prey.

Sharks aren't the only creatures with magnetoreceptor bumps on their noses, and they aren't the only underwater animals tuned in to magnetic fields.

---

* Before they bite, sharks often conduct a kind of taste test by bumping their noses into prey and emitting a short blast of electricity. If the signals conduct, as they do against animal or human flesh, there's a good chance the shark will bite. Wetsuits dull these signals, telling sharks, as Buyle says, "that we're not on the menu." Sharks assess the caloric value of their food on the first bite. If the prey doesn't register enough calories to justify a full-scale attack, the shark will release it and move on. Wetsuits might dull this sense and significantly decrease the chances of a return, full-scale attack seconds later.

In 2012, a group of German researchers were trying to figure out how trout could return to the same spawning ground every year. They suspected their ability to navigate blind underwater had something to do with the black bumps on their trout noses, which closely resemble the sharks' ampullae of Lorenzini. The researchers scraped off a few bumps and exposed them to a rotating electric field. The cells started spinning in sync with the field. In other words, trout had cells on their noses that worked the same way as a compass needle, and they were probably using these cells for navigation.

But perhaps the bigger discovery was that the bumps contained magnetite, a highly magnetic mineral that was used in early compasses.

Sharks, dolphins, some whales, and several other ocean migrators also have deposits of magnetite in their noses or elsewhere on their heads and are probably using them in the same way.

During full moons, some mollusks use magnetic north as a guide to move from deeper to shallower areas while they hunt. Even marine bacteria, which paleontologists believe date back over two billion years and may represent some of the Earth's earliest inhabitants, use tiny bits of magnetite to swim along magnetic field lines. This natural magnetic GPS has been around for billions of years, and, like all life, it began in the ocean.

Humans also have magnetite deposits. They're found in the skull, specifically in the ethmoid bone, which separates the nasal cavity from the brain. The location of these deposits in a human head corresponds closely to their position in sharks and other migratory animals—a relic from the magnetosensitive fish from which humans and sharks both evolved five hundred million years ago.

WHETHER OR NOT MODERN HUMANS can use the magnetite deposits or some other receptors to attune to the Earth's subtle magnetic field is still not known. But three decades of scientific trials suggest it's possible.

The first researcher to attempt to document and measure human magnetoreception was Robin Baker, a lecturer at the University of Manchester. Baker had long wondered how ancient Polynesian sailors could navigate hundreds of miles across open ocean and consistently find their way back home. Celestial or solar navigation could work some of the time, but not always — clouds covered the sky for days, and rough seas could quickly throw a boat off course.

Captain James Cook wrote about Tupaia, a high chief from Raiatea, near Tahiti, whom he took aboard his ship the *Endeavour* in 1769. Tupaia drew a detailed and accurate map that spanned more than twenty-five hundred miles, from the Marquesas to Fiji, and included 130 islands. For the next twenty months, the *Endeavour* sailed the South Pacific and beyond, and Tupaia could always point in the exact direction of his island home, regardless of the *Endeavour*'s location, the time of day, or the conditions at sea.

The Guugu Yimithirr, an Australian Aboriginal tribe, had a remarkable sense of direction that they incorporated into their language. Instead of using words meaning "right," "left," "front," and "back," Guugu Yimithirr used the cardinal directions of north, south, east, and west. If a Guugu Yimithirr tribesman wanted you to make room for him on a bed, he'd ask you to move a few feet west. Guugu Yimithirr didn't bend backward, they bent northward, or southward, or eastward.

The only way Guugu Yimithirr could communicate was by knowing their exact coordinates at all times, which was a hard thing to do at night or in an enclosed room. But it was second nature for them, as well as for a host of cultures throughout Indonesia, Mexico, Polynesia, and elsewhere, whose languages were also based on cardinal directions.

In the 1990s, researchers from the Max Planck Institute for Psycholinguistics, in the Netherlands, placed a speaker of Tzeltal — a Mayan directional language spoken by about 370,000 people in southern Mexico — in a dark house and spun him around blindfolded. They then asked the Tzeltal speaker (who was unnamed in the study) to point north, south, east, and then west. He did this successfully, and without hesitation, twenty times in a row.

The remarkable navigational abilities of these ancient cultures weren't exceptions; they were the norm. In a world without GPS and maps, knowing your exact location in a trackless desert, forest, or ocean was a matter of survival. All the people in these cultures developed an innate sense of direction that did not rely on visual cues. Robin Baker believed this sense was magnetic. In 1976, he decided to test it.

In Baker's first experiments, he blindfolded groups of students, drove them from the university along a winding route several miles out of town, and then led them, still blindfolded, one by one to an open field. He asked them to point in the direction of the university. The students frequently pointed in the right way, succeeding more often than pure chance would predict. He ran tests in different locations, at different times, with different students. In one test, the thirty-nine students pointed in the right direction with 80 percent accuracy, the same as closing your eyes, spinning around, and pointing to between 10:30 and 12:00 on a clock dial. Later tests had the same results. Baker repeated his experiment 940 times over the next two years with a total of 140 students. Overall, the experiments strongly suggested that the students were using some sort of nonvisual sense to orient themselves to their surroundings.

Baker then tested to see if the human navigational sense was magnetic. Earlier experiments with green turtles and birds had shown that tying magnets to the animals' heads would destroy their ability to navigate even very short distances. (The magnetic field from the head magnet was stronger than the Earth's magnetic field, and the theory was that it confused the animals into thinking that every direction they turned was north.)

Baker tied magnets to the heads of half the students, nonmagnetic brass bars to the heads of the other half, blindfolded them, drove a circuitous route out of town, and released each into an open field. The students without magnets were able to point in the right direction significantly more accurately than those with magnets. Additional tests yielded similar results. The magnets, Baker argued, were disrupting the students' ability to navigate, just as they did with birds and turtles.

After crunching the numbers from all his experiments, Baker wrote, "We have no alternative but to take seriously the possibility that Man has a magnetic sense of direction." The results were published in the prestigious journal *Science*.

Baker said that human magnetoreception was distinct from other senses, like vision and smell. Those senses are conscious, meaning that we are aware of them and immediately perceive it when they turn on (like when you open your eyes) and turn off (when you plug your ears).

Human magnetoreception works differently. It is an unconscious, latent sense; we can't feel it turning on or off in the same way that, most of the time, we don't notice that we're breathing. In this, magnetoreception is like the Master Switch; we don't know it exists unless we put ourselves in a situation in which we have to use it.

In the modern world, we're seldom given that opportunity. The pattern of settlements, roads, and other landmarks that undergird human society makes it easy to know our exact location at any time. As populations centralized, cities grew, and technology developed, the need for humans to have a keen sense of magnetoreception became dormant. Just like the need to hold one's breath and dive to gather food from the seafloor.

Baker's human magnetoreception results were met with fierce opposition. Dozens more experiments in human magnetoreception were attempted in the 1980s; some failed completely; others recorded mixed results. After ten years, however, the data was inarguable. The probability that the results of all experiments in human magnetoreception could occur by chance was less than .005. Statistically speaking, you'd have an equal chance of your house getting hit by lightning: one in two hundred.

For human magnetoreception to ever be proved, researchers needed to find out how it worked. They needed a receptor. In 2011, scientists at the University of Massachusetts medical school found one.

The researchers took fruit flies (which have a proven sense of magnetoreception) and removed a protein in their eyes that allowed them to sense and respond to magnetic fields. They then put the equivalent protein called hCRY2, from a human eye, and tested the flies' behavior. With the implanted human protein, the flies regained the ability to sense and respond to a magnetic field; the human protein in the eye has the same capacity to sense magnetic fields as the fruit flies.

Whether this protein is vestigial or actively being used in some sort of human magnetoreception is unclear. But Dr. Steven Reppert, the lead scientist of the study, said he would be very surprised if humans did not have a sense of magnetoreception. "It's used in a variety of other animals. I think that the issue is to figure out how we use it," he said.

For Robin Baker, the *CRY2* discovery was a vindication.

"I think one of the things that put people off accepting the reality of human magnetoreception twenty years ago was the lack of an obvious receptor," he said. "So these new results might actually be enough to tip the balance of credibility. I shall be fascinated to see."

IN THE END, IT ISN'T Buyle but seventy-four-year-old Guy Gazzo who tags Réunion's man-eating sharks.

After ten days of failed attempts tagging Réunion's bull sharks, Buyle heads back home to Brussels to pack up his camera gear for a documentary gig in the South Pacific. At his suggestion, Gazzo retools the spear guns to shoot at double strength and sails back out with Schnöller to Saint-Gilles. In one day, Gazzo tags three sharks, enough for an initial test of the SharkFriendly tracking system.

Schnöller and Gazzo spend the next month watching tagging data, trying to identify patterns. They notice an obsessive congregation around the Saint-Gilles marina. They decide to freedive into the area again, this time to investigate, not tag. The seafloor outside the Saint-Gilles marina immediately catches their attention. It's an enormous trash heap of plates, food, and refuse.

As it turns out, boaters at Saint-Gilles had been using the port entrance as a trash can. The bull sharks have gathered there to scavenge.

The humans who were attacked at the adjacent beaches had likely gotten in the way and were caught up in a localized feeding frenzy that was instigated by human activity.

The tagging discovery didn't scare people away; instead, it opened a new cottage industry. Tour operators began running shark-sightseeing snorkeling trips near the garbage dump. "We accomplished our goal," says Schnöller. By which he means, he's managed to educate people. Réunion's citizens now have a better understanding of the bull shark, its habits, and their role in violent exceptions to those habits.

Two months after the SharkFriendly campaign was initiated, the French government began reopening the beaches to the public.

# −800

A FEW MONTHS AFTER I visit Réunion, I'm back in Greece, sitting with about twenty other journalists on the patio of a restaurant in Amoudi, a bayside village on the southwestern edge of Santorini Island. We're waiting for a charter boat to take us three miles west, across the Aegean Sea to a bay near the island of Therasia. Out there is Herbert Nitsch, the self-proclaimed "deepest man on earth." In about an hour, Nitsch will attempt to ride a weighted sled to a depth of eight hundred feet on a single breath, what would be a world record in the no-limits discipline in competitive freediving and the deepest freedive ever attempted.

So far, however, things are not going well. The seas are rough and ocean currents are strong. Nitsch has never dived around Therasia and his team is worried that the currents might be powerful enough to bow the guide rope and slow his descent and ascent. Each wasted second decreases his chances of resurfacing conscious or alive.

The dive was supposed to happen at 11:00 a.m. It's now eleven and there's still no word on when the charter boat will be arriving. Some members of our group are threatening to leave; a few already have. Nitsch's main sponsor, the Swiss watchmaker Breitling, pulled out a few days ago. Nobody knows exactly why, and no one on Nitsch's team is talking, but the rumor is that executives at Breitling decided the dive was too dangerous.

The delay does nothing to ease my mind, specifically my mixed feelings about being here. My experience with Buyle a few months ago introduced me to freediving for a greater good. Freediving, I discovered, could be used as a tool to help crack the ocean's mysteries. It had a purpose.

No-limits diving was a step backward; it was another ego-driven competition, and one that put its athletes in great peril. I know this. And yet the part of me that loves superheroes, evolutionary leaps, and *Ripley's Believe It or Not* wants to see Nitsch travel to the outer limits of our amphibious abilities. I want to witness the deepest freedive ever attempted. And I'm not the only one.

Just three days earlier, a crew from *60 Minutes* landed on Santorini, along with lead anchor Bob Simon. He's sitting with the show's producer and a few cameramen at a table to my right. Simon plans to interview Nitsch right before the dive and witness the event alongside Nitsch's father on the team boat—the only journalist given such access.

That is, if the dive ever happens.

As the hours pass, Simon grows visibly annoyed. He pecks manically on his mobile phone and sips a Diet Coke. Someone at his table orders a plate of french fries; someone else behind me orders an iced tea. We stare at our mobile phones and wait.

Then, around noon, we hear an announcement. Nitsch's public relations manager, Silvie Ritt, instructs everyone to head for the dock at the north end of Amoudi Bay. The charter boat has arrived. We hastily pay restaurant bills, grab our bags, shuffle to the dock, and then hop aboard, filing onto bench seats of the exposed top deck. Outside, the wind still howls; waves crash on the seawall. The captain starts the motors and we head toward Therasia as gray swells chop and thump against the hull.

Everything seems to be falling apart, but the show, it appears, must go on.

NO-LIMITS, WHICH ALLOWS divers to use any means to attain depth, is the most extreme form of freediving and is, per capita, one of the deadliest sports in the world. Ten years ago, the no-limits freediving record was 525 feet. Since then, at least three divers attempting no-limits dives have died, and dozens have been injured, sometimes permanently.

In 2006, Venezuelan diver Carlos Coste returned to the surface paralyzed after a 597-foot no-limit attempt in Greece. Russian champion freediver Natalia Molchanova reported symptoms of brain damage after repeated no-limits training dives. In 2002, Benjamin Franz, a Belgian diver, resurfaced after a 542-foot dive totally paralyzed on his right side and unable to speak. He spent ten months in a wheelchair before he was able to walk, or swim, again. The list goes on.

The human body on its own cannot reach the depths of no-limits dives, which is part of what makes them so lethal. Most divers choose to strap into a weighted sled for the descent and then inflate an air balloon at depth to float them back up to the surface. These machines allow divers to plummet twice as deep as freedivers in other disciplines, usually in half the time. It's too fast for the body to eliminate the nitrogen in the blood that builds up during the deep descent. As a result, decompression sickness is a constant danger.

The sleds carry their own risks. Each is homemade, usually by the diver. Most freedivers aren't experienced with engineering watercraft. Case in point: Nitsch's sled was designed with the help of a twenty-eight-year-old whose day job is making prosthetic legs. It's the first sled of its kind. To gain depth, the sled will use a series of weights. When it gets to the end of the rope, an automatic trigger will release a blast of compressed air that will shoot it back to the surface. Or that's the idea.

The only way for Nitsch and other no-limits divers to test their sled designs is to try them during deep dives. They often malfunction. In October 2002, the French world champion freediver Audrey Mestre attempted to break the no-limits record with a 561-foot sled dive in the Dominican Republic. She made it down to her target depth, only to find that the air tank that was supposed to fill the balloon and buoy her back to the surface was empty. Many accused her husband, Francisco Ferreras-Rodriguez, of forgetting to fill the tank. Nobody else on board checked. Eight and a half minutes after Mestre began her dive, Ferreras pulled her body to the surface. Foam poured from Mestre's nose and mouth. She was blacked out but still registering a pulse. Without a proper doctor or even a stretcher on deck, rescuers propped up Mestre's body on a beach chair. She died shortly after.

SKETCHY MACHINERY, BLACKOUTS, and occasional deaths—all these things make no-limits dives almost unbearable to watch. And it's not like there's much to see anyway. As with other freediving disciplines, the action in no-limits dives happens below the surface. You see a diver huffing and puffing before the dive, see him take a final breath, and then, about four agonizing minutes later, you see him resurfacing—blue with asphyxia, often bloody. A trip to the emergency room usually follows. The whole thing looks insane.

Oddly, Nitsch himself seems anything but. When I met him at his hotel two days before the dive, I had trouble picking him out from his photographer, publicist, and other hangers-on. He is fit, taller than average, with a clean-shaven head, but not ripped with muscles or physically extraordinary in any other way. He speaks in the hushed monotone of a museum security guard, and he has lived, outside of freediving, a relatively mundane life in his native Austria, first as an airline pilot, then as an inspirational speaker for financial institutions. He appears completely, dreadfully, *normal*—and yet, it's this blandness about him, coupled with knowing the dangers of his profession, that give him a strange, almost-sadistic creepiness. A soft-voiced villain with hidden knives.

Nitsch started freediving "by accident," he told me, after an airline lost his scuba equipment on the way to a dive excursion in Egypt in 2000. Since then, he has broken thirty-two freediving world records in every discipline in the sport, and he has become the greatest overall competitive freediver in history.

His interest in going deep, he told me months ago when I first interviewed him on the phone, isn't about money or fame ("What money? What fame?" he asked) but about finding the human body's absolute limit, breaking it, and thus extending human potential. "If you think about what is impossible tomorrow," he said, "the day after tomorrow, you laugh about it."

BY THE TIME OUR CHARTER boat arrives at the Therasia coast, the wind has calmed somewhat and the sun is out, but the surface is still choppy, and the currents, I am told, are still strong. Nitsch's team is on a catamaran about three hundred feet to our north. On deck, a man is yelling. Crew members pace around, barking orders at nobody in particular. The screech of a mechanical winch cuts through the wind and the rumble of the boat motor. It's a chaotic scene.

Latched to the rope in the water beside the boat is Nitsch's sled, a black-and-yellow carbon-fiber pod that looks vaguely like a cough-suppressant gel cap. During his ascent, Nitsch will leave the sled at thirty feet below the surface and hold his breath for one minute, to allow nitrogen bubbles in his bloodstream to dissipate. The total dive time, Nitsch predicted, will be just over three minutes.

Neither Nitsch nor the scientists he's consulted know if he'll make it. If decompression sickness doesn't paralyze him, oxygen toxicity might. Most of what scientists know about the effects of oxygen on dives below 800 feet, they learned from physiologist Laurence Irving. For three decades, starting in the 1930s, Irving, who worked with Per Scholander, studied Weddell seals, which can hold their breath for up to eighty minutes and dive to depths below 2,400 feet.

The seals were also able to avoid decompression sickness by reflexively collapsing their alveoli, the small cavities that exchange gases in the lungs, at great depths. This collapse worked to minimize the uptake of air in the animals' bloodstream and prevent nitrogen from saturating blood and tissues.

It may be that alveoli collapse at great depths happens in humans. Nobody knows for sure, because no human has ever attempted to go as deep as Herbert Nitsch. The first step is for him to make the dive and live to tell about it.

NITSCH HAS EMERGED FROM THE cabin of the catamaran and is walking slowly around the deck. His head is bowed, and he's mumbling to himself. He steps down the ladder and enters the water. A diver hands him a float; Nitsch grabs it and leans his head back, so that he's facing the sun. Through his gaping mouth, he gulps air like a goldfish.

"Herbert Nitsch is about to begin the historic dive," a female voice announces through a loudspeaker on the charter boat. Nitsch pushes himself into the dive sled so that only his head remains above water. He is inhaling more deeply now.

"Get ready, everyone," the announcer squawks. A monitor on the catamaran calls out a two-minute warning. Nitsch's eyes are closed, his mouth drawing deeper breaths.

"Countdown," the monitor yells. Nitsch takes a big breath, then exhales. The monitor counts down from ten. Nitsch takes another huge breath, then exhales again.

"Eight . . . seven . . . six . . ." says the judge. The winch operator assumes his position behind a rack of levers on the back deck of the catamaran.

"Four . . . three . . . two."

By the time the judge reaches zero, the sled has disappeared below the surface.

"Twenty meters, thirty meters," says the judge, announcing Nitsch's depths from behind a sonar screen.

Nitsch's planned rate of descent is ten feet per second. In the first thirty seconds, he should be past three hundred feet, but he has reached only about two hundred. Something is wrong.

"Seventy meters, eighty meters."

"He's going too slow," I hear someone say behind me. A sickening tension mounts on the boat. Nobody moves.

"One hundred meters."

Forty-five seconds have passed. Nitsch should be down at around 450 feet, but he's a hundred feet short.

"One hundred twenty meters."

Ninety seconds pass and Nitsch is still sinking. At his current rate, he will be submerged for well over four minutes and will run out of air before he reaches the surface. He won't be able to pause below the surface to decompress, which will put him at greater risk of decompression sickness, oxygen toxicity, paralysis, and death. Meanwhile, the official at the sonar screen has stopped announcing depths. I ask the man next to me what's happening. "I don't like this," he says. "I don't like this at all."

Some two minutes later Nitsch's sled shoots to the surface. Nitsch himself is nowhere to be seen. Safety divers kick down. Nobody on deck moves or speaks. Thirty seconds later they return to the surface with Nitsch's unconscious body. His face and neck are bloated and bright red. A diver grabs an oxygen tank and mask from the catamaran and swims over toward Nitsch's limp form. Nitsch suddenly comes to.

"Give me the mask!" he yells, slurring his words. The safety divers don't know what to do; they stare blankly at each other. Nobody has trained for this kind of accident.

"Give me the mask!" Nitsch yells again. He is now barely breathing. He reaches a stiff arm over to the safety diver, grabs the oxygen tank and swim mask from his hands, then upends his body and tries to dive back down. He needs to give his body time to decompress. But he can't dive. Without weights, Nitsch's thick neoprene wetsuit buoys his body to the surface. He kicks his stiff limbs but doesn't seem to go anywhere. Each second wasted increases the chances of nitrogen bubbles entering his joints, lungs, and brain. Crew members on the catamaran stare at each other with wide, confused eyes, then watch helplessly as Nitsch flounders beneath them. The safety divers look at each other, look at Nitsch, and shake their heads.

"Let's have a round of applause for Herbert Nitsch!" the female voice announces through the loudspeaker. "The world's deepest man!" Someone claps. The rest of us silently gaze at Nitsch as he tries to kick his way down. Finally, he disappears. Minutes pass. Nobody knows where he's gone. We grimace and wait.

Five minutes later or so, the safety divers resurface, carrying Nitsch's body. He's blacked out again.

"Oxygen now!" a safety diver yells. They swim Nitsch to a waiting motorboat. He suddenly wakes up and tries to crawl onto the deck, but his arms buckle beneath him. The captain of the boat pulls him aboard and lays him face-up on the deck. Nitsch's eyes are puffy and swollen; veins bulge from his neck and forehead. He lifts his right arm and, with a shaking hand, points in the direction of Santorini. The captain guns the engine and the motorboat cuts a straight line to the hospital.

THAT NIGHT, NITSCH'S HEART STOPPED. Doctors resuscitated him and put him in an induced coma. The hospital staff shuttled him back and forth from a hospital bed to a recompression chamber, but their efforts came too late. Nitrogen bubbles entered his brain and cut off the blood supply to areas that controlled motor functions. He suffered half a dozen strokes. When he regained consciousness, days later, Nitsch could not walk, speak, or recognize his friends and family.

Later, I learn that the sled plummeted past its target depth, to 830 feet. He blacked out before he touched down, woke up during the ascent, and blacked out again at 330 feet. He was still in the sled, unconscious underwater, when the safety divers grabbed him at 30 feet and brought him to the surface. If they hadn't, Nitsch would have drowned. But as a result of the quick ascent,

Nitsch suffered a debilitating case of decompression sickness because he didn't have time to purge the nitrogen that had built up in his bloodstream.

Six months later he hadn't touched the ocean.

---

After the horrors of Nitsch's dive in Santorini, David King's near drowning, and Michal Rišian almost getting lost at sea, I swore off watching any more competitive freediving. Sure, the human body could dive deeper than scientists thought possible, but it also had limits. We all saw those limits. And I had gotten tired of seeing the bloody and blue faces of those who went beyond them.

In freediving, the ego is a deadly goad. It's also something of a blinder. Most of the competitive divers I met seemed to have little interest in exploring the deep ocean that they had painstakingly trained their bodies to enter. They dived with their eyes closed; nitrogen narcosis struck them dumb; they forgot where they were and why they were there. The deepest divers lolled themselves into a catatonic state that removed *any* sense of actually being in the water. The aim: Hitting a number on a rope. Beating your opponents. Winning a medal. Bragging rights.

Yes, they were swimming where no human had been before. But this struck me as maddening, like an explorer arriving in previously undiscovered wilderness and focusing only on his GPS coordinates.

This disconnect between athlete and ocean had me replaying the scenes in Santorini and Kalamata months after I'd returned home. My nightmares featured bloated necks and dead eyes. But my waking visions were more aspirational: I had seen Fred Buyle communing with sharks. His freediving took him to previously unknowable places, allowed him to see previously unseeable sights, and unlock hidden abilities. I too could access this world. The "doorway to the deep," he said, was open to everybody.

Those who'd reached the doorway described it in quasireligious terms — *transcendent, life-changing, purifying.* A new and shimmering universe. Getting there didn't require rupturing lungs or tearing a larynx. All you needed was a little training. All you needed was faith. All you needed was a certain level of comfort with voluntary asphyxiation.

And so, despite what I had seen, the more I thought about those freedivers, the more I wanted in. I wanted to flip the Master Switch.

NOBODY KNOWS THE MASTER SWITCH better than the ama, an ancient culture of Japanese diving women who once numbered in the thousands. For more than twenty-five hundred years, ama used the same freediving techniques, passed down from mother to daughter, to gather food from the ocean floor. In all written accounts of ama culture I found, there was never any mention of blackouts, bloodied faces, or drowning. Although they could dive up to a hundred and fifty feet down and stay under for minutes at a time, the ama never competed. Freediving for them was a tool, a means of survival. It was also a spiritual practice. The ama believed that when they approached the sea in their natural human form, they were returning balance to the world. "[Underwater] I hear the water coming into my body, I hear the sunlight penetrating the water," wrote one ama. They weren't visitors to the ocean; they were part of it.

Their story dates to 500 B.C., when shipwrecked nomads from central Asia found themselves stranded on the rocky shores of the Noto Peninsula, where there was very little vegetation and few land animals to hunt. The nomads turned to the sea, where they quickly adapted their bodies to harvest the bounty of life on the ocean floor. The women of the nomadic tribe — for unknown reasons, only the women — took over the daily diving and were later called the ama, which means "sea women." The ama didn't just survive their new aquatic lifestyle, they flourished, and soon spread along the Japanese coast and into Korea. Thousands, perhaps tens of thousands, of ama once lined the east coasts of the

Pacific and the Sea of Japan. By the 1800s they had become, in a sense, the world's largest commercial fishing fleet. European sailors who were lucky enough to see these half-naked diving women reported that they plummeted hundreds of feet on a single breath. Some claimed the ama could stay underwater for fifteen minutes at a time.

As fishing technology evolved in the nineteenth and twentieth centuries, the ama's numbers dwindled. Their villages disappeared. The daughters of the ama, who were to carry on the freediving tradition, left for more comfortable lives in the city. As of 2013, estimates of the total number of working ama ranged anywhere from a few hundred to zero.

According to the director of a short documentary film I saw online, a small group of ama were still working the waters outside of Nishina, a town 120 miles southeast of Tokyo in the rural Izu Prefecture. I wrote to various Japanese historians and tourist organizations but failed to find confirmation that the Nishina ama were still around. Nobody reported having seen them there for years. Nobody knew if they were still diving, or if they even existed.

A FEW WEEKS LATER, I fly to Tokyo, take a train to the Izu Prefecture, and rent a car in a beachside town called Shimoda. The locals in Shimoda repeat what my other sources told me, that I'm chasing a fantasy. They say the Nishina ama, who were supposed to live just ten miles up the coast, died off years ago. Or that the ama get in the water only a few times a year, mostly on holidays. Or that they are too old, frail, and tired to be visited. These anti-guides look at me with pity and point me down wet, dead-end streets. I follow their fingers for two days but come up empty. And then, on my third day driving the coast of Nishina, I come upon Sawada, a grungy little port full of broken boats and acrid smells. And I catch a break.

My guide is a lanky man named Takayan whom I found at the Nishina tourist office. He stands beside me on a breakwater, peering at a half a dozen divers bobbing up and down in the lead-gray water — what could be the last living link to one of the world's oldest surviving freediving cultures.

"Ama?" I ask Takayan again. I want to make sure we've found them. *"Hai,"* says Takayan. "Yes. Ama."

I turn and jog off along a gravel path to grab my recorder and camera from the rental car. When I return a few minutes later, the ama are dragging nets filled with their daily catch up the boulders of the breakwater and then stuffing them into duct-taped Styrofoam coolers.

"They are done for today," says Takayan. "They been diving all morning. We came too late." Behind Takayan, one ama, the shortest of the group, strips off her wetsuit pants and stands naked a few feet from me. When I walk away to give her some privacy, she chuckles and says something in Japanese to the three other ama standing nearby. They all laugh, then strip off their wetsuits.

Japanese society revolves around a tangled web of confusing customs. I figure I'm breaking a few dozen cardinal rules by approaching the ama without an invitation (or without offering a gift, or without speaking Japanese, or simply by being male). But I've traveled seven thousand miles on a whim, based on a home movie I'd seen on the Internet. After many false starts, I'd found them. And there was a good chance I'd never find or speak to them again.

I walk to the water's edge and cringe for a few minutes, waiting for the ama to change into their sweatpants and tattered rain jackets. Then I reapproach, smiling. The ama don't smile back.

"They are too tired," Takayan says, stopping me. "They don't want to talk now." He says I'll have a better chance if we come back tomorrow at dawn, before their morning dive. What he's really saying, I think, is that the ama want to see how dedicated I am. Returning early tomorrow will show them I'm truly interested in their culture, that I'm not just a tourist wanting a quick look.

I walk to my car and watch through a dirty windshield as the ama pack their gear into rusting shopping carts. I am still watching, minutes later, as they file slowly past the broken boats and empty lots of Sawada and then vanish into white fog.

THE AMA MIGHT HAVE BEEN the largest group of freedivers in history, but they weren't the first. Archaeological evidence of ancient freediving cultures goes back as far as ten thousand years. Written accounts of freedivers date to 2500 B.C. and span the Pacific, Atlantic, and Indian Oceans.

Around 700 B.C., Homer wrote of divers who latched themselves to heavy rocks and plunged below one hundred feet to cut sponges from the seafloor. In the first century B.C., trade between the Mediterranean coast and Asia exploded, in part because of red coral, a favorite cure-all in Chinese and Indian medicine. Most red coral grew at depths below a hundred feet and could be collected only by freediving. By the eighth century, Vikings in the North Sea were freediving beneath enemy ships and boring holes in their hulls to sink them.

Then there were the pearl divers, who flourished in the Caribbean, South Pacific, Persian Gulf, and Asia for more than three thousand years. When Marco Polo visited Ceylon (now Sri Lanka) in the late fourteenth century, he witnessed pearl divers plummeting more than a hundred and twenty feet on dives that lasted from three to four minutes.

In 1534, Gonzalo Fernández de Oviedo, a Spanish historian visiting Margarita Island in the Caribbean, observed indigenous Lucayan Indians descend to more than a hundred feet on dives that, according to his notes, lasted fifteen minutes.*

These weren't just champion divers, either. According to Oviedo, hundreds of Lucayan shared this incredible breath-holding ability, which they used to dive deep from sunrise to sunset, seven days a week, without ever appearing to tire.

Within a few years, the entire population of freediving Lucayan had died of disease or been enslaved and shipped off to

---

* Some sources claim that Oviedo intended to write 5 minutes instead of 15; others contend his reporting was accurate.

gather pearls on other islands. The Spanish brought in African slaves to take their place. The Africans took to freediving almost immediately and, according to reports, were soon diving to a hundred feet, holding their breath underwater for as long as fifteen minutes.

Halfway around the world, in Indonesia, Sir Philiberto Vernatti, a field scientist who worked with one of Britain's esteemed scientific organizations, the Royal Society, reported in 1669 that he saw pearl divers stay underwater for "about a quarter of an hour." Similar accounts of fifteen-minute dives came in from Japan, Java, and elsewhere.

Which wasn't to say these long dives were easy. According to the same reports, once at the surface, many divers suffered from violent seizures: water and blood poured from their mouths, ears, nostrils, and eyes. They'd sit and recover for a few minutes, take a deep breath, and do it all over again. Some dove forty to fifty times a day.

All told, there were a dozen accounts from unrelated travelers to different locations over the course of centuries that described the same thing: dives to one hundred feet, with divers lasting up to fifteen minutes on a single breath. And there was never a mention of air tubes, special diets, or metabolism-depressing drugs to assist these freedivers. In fact, most of the Caribbean divers lived under lock and key in deplorable conditions and would puff pipes or cigarettes between dives, sometimes in the water just before heading down.

And then, poof. Gone. By the twentieth century, pearl farming and new fishing technologies had made freediving obsolete. The human body's amazing diving abilities and human knowledge of freediving began to disappear. A person like me, who has spent decades swimming in the ocean, wouldn't think of holding his breath for more than thirty seconds on purpose.

Today, modern competitive divers are rediscovering the ability, but they aren't as good at holding their breath, if historical reports can be trusted. Did these long-ago cultures know something ours doesn't? Are there some ancient Japanese secrets to breath-hold diving that could help me hold my breath longer and dive deeper? Are we just now rediscovering our true potential in water?

If anyone could tell me, it would be the ama.

AT DAWN THE NEXT DAY, Takayan and I return to Sawada to find four ama sitting in a circle on a patch of concrete above the breakwater. They are drinking green tea and eating dried seaweed and yogurt, joking with one another, sometimes laughing so hard that they spit up their food. Hardly a moment goes by without one of them throwing her head back and cackling into the air.

The ama are the opposite of the demure, meticulously styled women I had seen elsewhere in Japan, and nothing like the fairy-tale ama advertised in movies, old woodcuts, and turn-of-the-century daguerreotypes.

This group is bawdy, brazen, and gruff; their skin is tawny and wrinkled from decades of salt water and sun. They have unkempt hair and wear ripped clothes. In short, they are a refreshingly motley crew who don't really seem to give a damn what I or anyone else thinks of them.

Takayan exchanges a few words of Japanese with them, the ama nod, and then he introduces me to the group.

Yoshiko, who is tall with reddish hair and a long face, is sixty and has been diving since she was eighteen. The other two ama are at least ten years older and have been diving since they were fifteen. Although these three women aren't related, they share the same last name (Suzuki) and claim to be descended from a centuries-long lineage of freedivers. Another, smaller woman, with frizzy hair like a Chia pet, says she is eighty-two. Her name is Fukuyo Manusanke and she's been diving since she was in her thirties. She's the loudest of the bunch.

With Takayan translating, I ask Manusanke some questions. I'm told that the ama have always been made up of women — not because they were subservient to the men, as many historical books claimed, but because only women understood the rhythms of the sea. Manusanke points to the commercial fishing trawlers motoring out to sea from the Sawada port. These boats, with enormous nets tied to their sides, indiscriminately pull up whatever is nearby. Many of the fish, jellyfish, and other animals the trawlers catch are unusable; their corpses will be thrown back as trash. Manusanke says the fishermen on these boats destroy the environment and ruin the natural balance of the ocean.

"When a man comes to the ocean, he exploits it and strips it," she says. When a woman puts her hands in the ocean, that balance is restored. Manusanke explains that the ocean can always provide for humans if they gather from it in their natural forms. A person should take what he or she can carry, but no more. Otherwise, she says, eventually, there will be nothing left.

Until just sixty years ago, the ama didn't even wear swim goggles, fearing that this would allow them to see too much and give them an unfair advantage over other creatures in the sea. They didn't use wetsuits until the 1980s. Some ama still dive topless.

There were once sixty ama around Nishina. Manusanke says in the past twenty years or so, the number has dwindled to twenty-five. The few ama left don't dive often. The twenty-five-hundred-year-old line of ama tradition is about to be broken. "We are the last ones left," she says.

The ama excuse themselves and walk over to the beat-up shopping carts holding their wetsuits and dive gear. The interview is over, they tell me. It's time to go diving.

I brought my gear from San Francisco, hoping to make a dive and see the ama's ancient breath-hold techniques in action. The ama don't seem thrilled about the prospect, but they agree to bring me along for a few hours. I run back to my car, grab my stuff, and join Manusanke and the others near water's edge.

While the ama pull on faded, torn scuba-diving suits, I slip into a four-hundred-dollar freediving suit I'd just had custom-made in Italy. As they defog antique masks with local yogumi leaves, I squirt chemical-packed liquid defogger on my new low-volume goggles. While they slip on castoff bright yellow boogie-board fins, I wiggle my feet into state-of-the-art, three-foot-long, super-efficient camouflage freediving fins.

Manusanke points at my fins and guffaws. Yoshiko taps on the glass of my mask and shakes her head. An ama named Toshie Suzuki, who sports an unruly nimbus of curly hair, touches my suit, and then takes her finger away quickly and shakes it, as if she has just come into contact with something contaminated. I feel like an idiot.

And then I begin to see why the ama have kept the ocean to themselves all those centuries, and why they're apprehensive about sharing their secrets with outsiders, especially men. There I am, the typical male, exploiting the latest technology to find a shortcut into a world I only dimly understand. In some ways, I'm no different than the fishermen on the trawlers sailing out behind us. I am disrupting the balance of the ocean that the ama have spent the last twenty-five hundred years trying to protect.

Manusanke and the rest of the ama hobble down over the boulders of the breakwater and splash into the water. They are laughing, yelling at one another, and making squeaking sounds like dolphins. I follow and we swim out together toward the horizon, until the Sawada port fades into the fog. The Suzukis head east, just outside the rocky cliffs of the bay, while Manusanke and I stay put. I watch as she adjusts her mask, takes a deep breath, makes a whistle that announces to the other ama that she's diving, then flips upside down and goes. She kicks her eighty-two-year-old body down past five feet, then ten feet, then twenty feet, and deeper, until her movements soften, she stops kicking entirely, and she descends effortlessly, dissolving into the black water below.

I take a breath and try to join Manusanke on her dives, but even with all my state-of-the-art equipment, I'm buoyed to the surface. My deepest dives are about twelve feet; my longest last twenty seconds. Any more, and I start getting claustrophobic and panicky, and the ache in my ears and head becomes excruciating. The harder I try, the more painful the dives become. I eventually give up.

• • •

BY NOON, WE'RE BACK ONSHORE, sitting in a half circle along the breakwater. The ama have emptied their fishing nets on the concrete in front of us. They've each gathered a few dozen sea urchins, which they'll sell for a pittance to nearby sushi restaurants. I gather my gear and thank the ama, then Takayan exchanges a few words of Japanese with Manusanke and the other women. They laugh, then wave goodbye, smiling.

As Takayan and I walk back to our cars, I ask him what was so funny. He says that he asked Manusanke if the ama had any ancient freediving secrets they'd like to share with me.

"And? What did she say?"

" 'You just dive,' " Manusanke had told him. " 'You just get in the water!' "

It was the same answer I got from Otto Rutten at Aquarius, the same answer Fred Buyle, Hanli Prinsloo, and the other divers in Greece had given me. There was no shortcut, no rulebook, no secret handshake, no specialized equipment or diet or pill that would get me there. The secret to going deep, they all seemed to be saying, was within each of us. We're born with it.

But unlocking that secret was trickier than I ever imagined.

# −1,000

"IT WAS LIKE A NEAR-DEATH experience. Like being transported to some other place, some other plane," says Fabrice Schnöller.

Schnöller and I are sitting together at Planet Nature, a health-food restaurant he owns with his wife in downtown Saint-Denis, Réunion. Schnöller uses the second-floor seating area as his office, but it looks more like a storage room. USB cables and electrical cords cover a desk like ivy. Stacks of scientific papers teeter on the tables. Rows of dog-eared academic books cram corner shelves.

It's a few months after my meeting with the ama, and, against the complaints of my lower back, I've returned to Réunion. I've been lured here by Schnöller, who e-mailed me a few weeks ago and said he was on the cusp of a "big discovery." It had something to do with whale and dolphin clicks, but he was short on specifics. He told me he'd invited a team of scientists, researchers, and freedivers from all over the world to come to Réunion for a week-long conference to discuss it.

"You should join the team," he said. I accepted and flew thirty-two hours to reconnect with him and his crew. My plan was to stay ten days.

Now, a few hours before the conference convenes, Schnöller sits me down and recounts how he sold his business and decided to dedicate his life to studying dolphin and whale click communication.

"It started while I was sailing to Mauritius, about five years ago," he says, taking a sip of Dodo beer. "That's when everything changed."

SCHNÖLLER WAS CAPTAINING a sixty-foot sloop called the *Annabelle* that his friend Luke had just bought from Paloma Picasso, a daughter of Pablo Picasso (an original Picasso still hung in the hull). A few hours into the thirty-six-hour trip, Luke and the six other crew members were below deck incapacitated by seasickness. The onboard chores fell to Schnöller.

Schnöller didn't mind. He enjoyed captaining the *Annabelle*, mostly because he liked being alone at sea, especially in the dark. Around eleven on the first night, he leaned back in the captain's chair and stared out across a gaping sky, glistening with stars. In his left hand, he held a thermos filled with coffee; with his right, he turned the oversize wheel of the ship to the northeast. He listened to the syncopated thump of waves crashing on the prow, and he imagined that the sound was coming from some large hand reaching up and tapping out a pattern on the ship's bottom, like fingers on a bongo drum. Through his headphones, the spiraling bass line of the Doors' classic "Riders on the Storm" faded in; the sound of canned winds and rain from the song mixed with the real wind and splatter of spindrift that blew salty water on his face and hair. Schnöller smiled, sailed on, and watched as the blackness of night drained out of the sky like water from a dirty sink, leaving only clear blue and orange in its wake. It was morning again.

By 10:00 a.m., the winds had calmed and swells subsided. The crew members slowly emerged from the cabin with puffy eyes and swollen faces, exhausted. Captain Luke apologized for leaving Schnöller on watch all night. Schnöller nodded, took another sip of coffee, and swallowed the last bite of the sandwich Luke had packed for himself to eat later. He kept his eyes on the horizon. Luke noticed a column of mist at the side of the boat — it looked as if a grenade had detonated in the water. Then another little bomb went off, and another. Schnöller had heard from other sailors that there were whales in this stretch of the Indian Ocean. Seeing them at a distance was common, but having your ship surrounded by them was practically unheard of. Schnöller felt a strong urge to jump in and swim with them.

He went below to grab his mask, fins, snorkel, and a waterproof camera. Luke met him at the back of the boat and begged him to stay on deck. Another crew member, Jean-Marc, joined Schnöller on the transom, and they both jumped in the water.

The ocean is usually silent, but the waters here were thundering with an incessant *click-click-click,* as if a thousand stove lighters were being triggered over and over again. Schnöller figured the noise must be coming from some mechanism on the ship. He swam farther away from the boat, but the clicking only got louder. He'd never heard a sound like this before and had no idea where it was coming from. Then he looked down.

A pod of whales, their bodies oriented vertically, like obelisks, surrounded him on all sides and stared up with wide eyes. They swam toward the surface, clicking louder and louder as they approached. They gathered around Schnöller and rubbed against him, face to face. Schnöller could feel the clicks penetrating his flesh and vibrating through his bones, his chest cavity.

"I felt like I was contacting ET, you know, like this was communication from another planet," says Schnöller. He and Jean-Marie spent two hours swimming with the whales that day. He'd known nothing about whales before the encounter. Afterward, they became his obsession.

When Schnöller got home to Réunion, he Googled images of whales and compared the photographs he'd taken with those he found online. Sperm whales are the largest of the toothed whales, and, according to historical accounts, the most ferocious whale predators. Historical images that Schnöller found depicted them killing humans, crushing boats, gorging on giant squids. But in his brief experience with them, Schnöller saw the whales to be gentle, curious, and intelligent, making him wonder how accurate those old images were. With their eight-inch teeth, the whales could easily have killed him. But instead they approached in peace and welcomed him into their pod. Schnöller wanted to understand how history and reality could be so far apart. He looked around for the latest sperm whale–behavior studies. But there weren't any.

"I assumed the military, thousands of scientists around the world, were conducting studies on these animals," he says. "I found nothing—no research, no videos, no photographs."

Schnöller realized the only way he could get his wife, the other crew members on the boat, or anyone else to understand the experience he had was to begin his own research program. Six months after the sperm whale encounter, he sold his lumber supply store and started the nonprofit DareWin. He enrolled in biology courses at the University of Réunion. He learned that dolphins, beluga whales, orcas, and other cetaceans also use the distinct clicking sounds he'd heard and felt swimming with sperm whales.

Sperm whales rarely visit Réunion's coast, but bottle-nosed dolphins, which can dive down to a thousand feet, are common. Schnöller focused on recording interactions between dolphins and analyzing their vocalizations, which include clicks, burst pulses, and whistles.

Over the past five years, Schnöller and his small crew of volunteers recorded more than a hundred hours of wild dolphin behavior—the largest collection of its kind in the world.

SCHNÖLLER GETS UP AND LEADS me to his desk. Behind a lump of papers is an oversize computer monitor with a spectrogram readout—a visual representation of an audio signal—showing some dolphin clicks and other vocalizations he had recorded months earlier. He starts a track, which he says contains rapid click patterns called burst pulses. The speakers blast what sounds like party whistles and machine-gun fire. "All those noises, they are all coming from one dolphin," he says. *"One dolphin."*

Dolphins and other cetaceans use these clicking sounds as part of a sophisticated form of sonar called echolocation. They're similar to the clicks sperm whales used to shake Schnöller's body years ago, only weaker.

To understand cetacean echolocation, Schnöller says, you first need to understand sonar.

A simple sonar system, consisting of one speaker and one hydrophone (an underwater microphone), works by first sending out a pulse sound, or ping. That ping travels through water until it hits something, then echoes back. The hydrophone records the echo, and a processor calculates how long it took for the echo of the ping to return. This system can provide information on how far away an object is and the direction it is moving, but nothing more.

A more complex sonar system includes dozens of hydrophones distributed over a wide area. When a ping is sent out, the echo that returns reaches each of these hydrophones at a slightly different time. With this extra information, the sonar system can determine not only the distance of an object, but its shape and depth. A rough picture emerges.

Dolphins and some whales have the equivalent of thousands, even tens of thousands, of echo-collecting hydrophones built into their heads. When a cetacean sends out a click (its version of a sonar ping), it receives the echo information with a fatty sac located beneath the lower jaw. Unlike ears, which provide only two directional sources to gather information, this fatty sac provides the cetacean with thousands of data points. The animal can process these to gauge the distance, shape, depth, interior, and exterior of the objects and creatures around it.

Dolphins can detect the shape, position, and size of larger objects from up to six miles away. Their echolocation is so powerful and sensitive that it can penetrate over a foot deep into sand;

it can even "see" beneath skin. Dolphins can peer into the lungs, stomachs, and brains of the animals around them. With all this information, scientists believe dolphins can create the equivalent of an HD-quality rendering of objects nearby — not only where these objects are, but how they look from the inside out. In essence, dolphins and other cetaceans have x-ray vision.

Echolocation isn't just a curiosity; it's essential for a cetacean's survival. Ninety percent of the ocean is cloaked in permanent blackness, and even those areas near the surface are black at night. To adapt to this dim-lit environment, some animals evolved to have super-sensitive eyes; others create their own light with bioluminescence; rays and sharks use electro- and magnetoreception. Cetaceans evolved to have remarkable powers of echolocation.

THIS "SENSE" IS NOT RESTRICTED to the ocean. Bats have used echolocation for fifty million years to thrive in complete darkness. Humans have echolocated for hundreds, perhaps thousands of years.

French philosopher Denis Diderot noted instances of "blind sight" in the mid-1700s. Almost a century later, in the 1820s, a blind English adventurer named James Holman traveled around the world using a self-taught form of echolocation. The public was skeptical about Holman and other human echolocators. Most people believed they had partial vision or perhaps were using something called facial vision, the sensation of pressure increasing on one's face as one moves closer to an object. In 1941, a Cornell University psychologist named Karl Dallenbach tested for it.

He gathered a group of blind subjects and had them walk toward a wall. When they believed that they could perceive the wall, they were told to raise their left arms; when they thought they were about to collide with the wall, they were to raise their right arms. The blind subjects succeeded in perceiving the wall from dozens of feet away; they stopped mere inches before walking into it. Dallenbach then gathered a group of sighted subjects, blindfolded them, and repeated the experiment. The sighted people perceived the wall almost as accurately as the blind ones.

DEEP

Next, Dallenbach had the subjects walk on a carpeted path that an assistant blocked with a board at random intervals. After thirty trials, the blindfolded, sighted subjects could locate the board just as reliably as the blind subjects. Dallenbach then tested for facial vision by putting felt hoods over subjects' heads, which would minimize any pressure sensations they could pick up from their surroundings. The hooded group easily perceived the board and wall as accurately as they had without the hoods. Dallenbach concluded, correctly, that humans did not use facial vision; we too had a sixth sense of echolocation.

A FEW WEEKS AFTER SCHNÖLLER introduced me to the otherworldly concept of echolocation, I'm walking down a street in suburban Los Angeles with Brian Bushway, one of the world's most gifted human echolocators. As we stroll toward a restaurant he's chosen for lunch, Bushway emits a sharp, short click from his mouth, then points out an empty driveway on our right, a van parked to our left, and rows of overgrown bushes on an upcoming corner. He clicks in the other direction and mentions that the house we just passed is small and covered in plaster, while the one across the street has large bay windows. The lawn in front of the apartment complex just ahead is in sore need of gardening. Bushway gets to the end of the sidewalk, pauses a moment, then leads me past two parked cars and up a curb to the sidewalk on the other side of the street. We take a right, he clicks again, and then he walks me through a crowded parking lot. He tells me the Cuban restaurant we're going to is up this way. I follow him through the entrance and into a crowded dining room. A server walks us to a corner table and hands us menus. Bushway puts the menu down, without looking, and tells me to order for him. He can't read the menu; he can't even see it. He's blind.

I learn about Bushway through YouTube videos. I watched him dashing along a dirt trail on a mountain bike, nimbly dodging branches, bushes, and boulders, and then riding down a steep flight of stairs. Next, he was jogging across a river and through

104

a thirty-foot-wide mud pit. Another video showed him walking through a park, approaching a tree, and climbing it.

Bushway, who has a muscular frame and a frizzy mop of hair, tells me he began losing his vision at fourteen. One day, he couldn't make out writing on a school chalkboard. A few weeks later, while playing hockey, he couldn't find the puck. He started having trouble recognizing his friends. A new set of contact lenses didn't help. When he woke up one morning and saw that everything in his field of vision was bright white, his mother rushed him to the hospital. A doctor dilated Bushway's pupils, then turned off the light to conduct a routine check.

"The lights never came back on again," says Bushway, taking a napkin from the table and placing it on his lap. "After that, I remember walking out of the office with my mom and asking, 'Is the sun out?'"

The sun was out, but for the first time in his life, Bushway couldn't see it. He would never see anything again.

Bushway suffered from optic nerve atrophy, a rare disease that destroyed the optic nerves in both eyes. After he got home from the doctor's office, he spent the next several months feeling helpless. Doctors recommended doing a biopsy of his optic nerve so they could test whether the damage was genetic. Surgeons shaved half of his head, cut out a section of his skull, moved his brain aside, and clipped out part of his optic nerve. After the surgery, scar tissue formed on his brain. He started having seizures. Doctors put him on antiseizure medication, which gave him extreme vertigo and constant shakes. "I felt uncomfortable moving," he says. "I just sat on the couch and listened to talk radio and books on tape." The highlight of his day was going with his mother to a drive-through restaurant, picking up food, coming home, and eating it.

Bushway returned to school a few months after he lost his sight. In the past, he had prized his independence and lived an active lifestyle. Now, an adult needed to guide him around campus. He could no longer play sports, walk by himself, or relate to his friends. He felt like an outcast, totally alone. He dreaded the idea of living the rest of his life this way.

Weeks later, while standing in the courtyard of his school, he suddenly sensed something in front of him. It was a pillar. He noticed several more pillars next to it. "I wasn't touching them," he says. "I was five feet away, but I *swore* I could see them. I could count them — it was like a sixth sense; it's even a magic power."

Bushway was soon back to riding his skateboard, shooting hoops, and roller-blading. He joined a mountain-biking team and bombed down local trails. His vision hadn't returned; the optic nerve damage was irreversible. Instead, another sense within him had suddenly turned on that allowed him to "see" through his blindness. With this sense, Bushway could point out a car in a parking lot a hundred yards away, tell you the width of a tree trunk from across a sidewalk, and distinguish a Rubik's Cube and a tennis ball from across a dining-room table.

He honed these skills with the help of a blind activist named Daniel Kish, whom he'd met at a lunch for blind students a few weeks after first sensing the pillars at school. Kish, who had lost his own sight at the age of one, ran a nonprofit organization called World Access for the Blind. The program taught blind people how to use an echolocation system that Kish had developed called FlashSonar.

FlashSonar isn't a device; all the tools required to use it exist inside the human body. And the "magic power" that allowed Bushway to first see the columns in the courtyard at school wasn't magic at all, Kish explained. It was the same echolocation sense that dolphins and whales used to navigate through dark ocean depths for the past fifty million years. Humans could also "see" in the dark, he said. Most of us had just forgotten how.

BACK AT THE CUBAN RESTAURANT, I watch Bushway emit a quick, crisp click from his mouth, pause a second, and then reach across our table to grab a water glass. We pay the bill and Bushway clicks again as we get up from our table. He's still clicking when he leads me out of the crowded restaurant, through a parking lot, and across bustling sidewalks. At the footpath to his apartment building, he stops, tells me to watch my step, and then takes me through his front door.

It's time for my first lesson in FlashSonar. He asks me to stand arm to arm with him in the middle of his living room. He lifts his tongue to the roof of his mouth and slaps it down just behind his bottom teeth, releasing a click. He listens for the echo of this click to determine the shape and distance of things around him.

For instance, a wall three feet from him will reflect an echo faster than one further away. Objects sound different too, depending on their structure and materials. "If something looks soft," Bushway tells me, "it will *sound* soft." A wooden wall, for instance, absorbs more sound, so the echo will be more muted than that of a glass door. Bushway perceives these differences almost instantly.*

He clicks, then walks across his apartment living room and enters the kitchen. He bends down, opens a drawer, and pulls out a cutting board. He clicks again, approaches within two feet of me, stops, puts the cutting board at an arm's length to the left of my head, and ties a blindfold over my eyes.

"Now click," he says. I slap the tip of my tongue down to create a popping sound. With my eyes still blindfolded, I hear Bushway walk to my right side. He holds the cutting board up (although I can't see it) and tells me to click again. I immediately sense a difference in the echoes. Within a few minutes I can identify the location of the cutting board at different spots around the room from a distance of about six feet.

I take the blindfold off. I'm feeling pretty confident, but Bushway jokingly tells me not to get too excited. Five-year-olds can do what I'm doing, and can probably do it better.

He mentions a Spanish study in which researchers took ten sighted volunteers and taught them the basics of FlashSonar dur-

* By the time a click leaves a human echolocator's mouth—traveling at 1,100 feet per second—and returns to his ears and creates an image in his brain, a few thousandths of a second have passed.

ing two training sessions. Each session lasted an hour or less. Afterward, the students were placed in an empty fifty-foot-by-fifty-foot room. A stereo played white noise and complex echo patterns in the background to mimic a real-world environment. Volunteers were able to detect flat surfaces like walls, wooden panels, and flat monitors from about thirty feet away. As they walked around, they were able to stop twenty inches away from hitting walls.

In 2011, a team of Canadian researchers placed Kish and another blind echolocator into fMRI machines and recorded the activity in their brains as they used FlashSonar techniques. The researchers then brought in two sighted subjects who had never used FlashSonar and had them click at their surroundings while they were being scanned by the fMRI machines. They compared the scans of the blind FlashSonar users with those of the sighted subjects. The scans revealed that when Kish and the other blind echolocators used FlashSonar, the visual part of their cortexes lit up. Sighted people showed no activity in this area when they clicked.

These findings suggested that FlashSonar users were processing auditory information in much the way the rest of us process visual information. The echolocators were, in essence, seeing via echoes.

The Master Switch and magnetoreception are latent and unconscious senses. We never know they are working. Human echolocation, however, is patent—we can consciously *hear* its effects and "see" its effect. And with some practice, anyone with decent hearing can hone this nonvisual sense of sight.

Bushway now works with Kish as an instructor for World Access for the Blind. In the past five years he has helped teach FlashSonar to more than five hundred blind people in fourteen countries. "When you go blind, the blind community gives you a cane, a dog, shows you how to go to the post office and a restaurant, then you come home," he says. FlashSonar is a way to regain total freedom.

He tells me to put the blindfold back on. Then he opens the front door and leads me into a world as black as the ocean's lowest depths. I stand still, waiting for my ears to acclimate to the sounds of the night city. Slowly, LA comes into focus in a new way, sounding sharper and richer than I've ever heard it before.

"Now," Bushway says, "click."

LIKE CETACEANS, WE TOO CAN use clicks and echoes to perceive and navigate through our world. Fabrice Schnöller believes cetaceans also use these sounds to communicate with one another.

Back in Réunion, he closes the dolphin burst-pulse file on his computer and opens another audio file. The echolocation discussion is over, he says. He now wants to tell me why he's invited me and a group of scientists, freedivers, and researchers to come here for the week. It concerns cetacean clicks, he says, but has nothing to do with seeing in the dark.

"I want you to look at this." He points at the computer screen in his unkempt office. "Look at how coordinated it is." On the display are two spectrogram readouts of dolphin vocalizations called whistles. The whistle patterns are precise, each separated from the next by the exact same millisecond-long interval.

Schnöller believes that cetacean clicks and whistles underlie a sophisticated form of communication. He plays two more dolphin whistles whose spectrogram patterns look identical to the last two. Dolphins can repeat these whistles in the exact same frequency and length over and over again. They can then add slight variations, repeat these multiple times, change them slightly, and so on. Schnöller says each of these whistle patterns could represent some form of language. "This isn't, you know, your dog barking." He laughs.

In one of his first experiments, in 2008, Schnöller downloaded dolphin whistles to a waterproof mobile phone and headed out in a motorboat along Réunion's coast with his twelve-year-old daughter, Morgane. An hour later, dolphins approached the boat. Schnöller took an underwater video camera while Morgane grabbed the mobile phone, and the two jumped in. When they got within a few feet of the dolphins, Morgane pressed the phone's Play button.

"It was the same as a dolphin popping its head out of the water and saying, 'Hello, James,'" Schnöller explains. "Only I'm not sure what exactly we were saying to him. We could have been saying hello, or we could have been telling him to fuck off!"

One dolphin in the receiving pod, whom Schnöller named QuackQuack, stopped suddenly, did a double take, and replied with a series of high-pitched whistles, then swam away. Morgane turned up the phone's volume and pressed Play again. Quack-Quack stopped, turned, and repeated his reply.

"He thought we were really talking to him," says Schnöller. "Like we had learned their language or something!"

In the months that followed, when Schnöller went to sea, QuackQuack would often find his boat, approach, and start vocalizing, as if he were picking up the conversation where they'd left off.

Schnöller tells me dolphins use specific, extremely detailed signature whistles to identify themselves in large groups. A mother dolphin will often whistle the same pattern to a newborn for days — a way, some marine biologists believe, to imprint a name on the baby. Dolphins use these name signatures when they approach other dolphins, to identify themselves. They also speak their names when they approach humans. Schnöller reasons that when QuackQuack heard a whistle blast from the mobile phone, he immediately replied with his name. He was introducing himself.

Last year, Schnöller created his own signature whistle, essentially his own dolphin name, to introduce himself to the dolphins. He specifically corrupted the whistle so that he could distinguish it from other dolphin whistles should the dolphins learn it and speak it back to him. All dolphin whistles ever recorded have been in the form of smooth sound waves. Schnöller's whistle was very harsh in acoustic terms, a sharply angular square waveform

—a form no dolphin had ever been recorded using. He motored out to the coast, tracked down a pod, got in the water, and started playing his strange signature whistle.

"The first time we try it they are very, very interested, but did not make any imitation," Schnöller tells me. Six months later, Schnöller was back in the water recording the whistles of a different pod of dolphins. When he returned to his office and analyzed the recordings, he discovered that all ten dolphins in the pod had adopted his square signature whistle form into their whistles.

"They were using it in their language!" Schnöller says. To him, this was like traveling to some distant village in China to find that everyone knew your name.

Cetaceans have disproportionately large and complex brains compared with other animals. The brain of the bottle-nosed dolphin, for instance, is about 10 percent larger than that of a human, and in many ways more complex. For instance, the dolphin neocortex, the part of the brain that performs higher-order thinking functions like problem-solving, is proportionally larger than the human neocortex. To Schnöller, who had spent months in a brain lab while in college, this was no coincidence. It proved to him that dolphins and other cetaceans were very intelligent and capable of sophisticated communication.

Dolphins don't have vocal cords or larynxes, so they can't vocalize in a way that sounds like human speech. Instead, they use two small mouth-like structures embedded in their heads—vestiges of what were once nostrils. The dolphin can flex and bend these nasal passages, called phonic lips, to create a variety of sounds—whistles, burst pulses, clicks, and more—in frequencies that range between 75 and 150,000 Hz. Scientists did not detect many of these sounds for years because humans can't hear them. (Normal speech of an adult human is between about 85 and 300 Hz, although we can vocalize through consonants and overtones up to about 20,000 Hz.) The only way scientists discovered the dolphins were communicating at such high frequencies was by recording them and then playing the sounds back through a spectrogram. When they did, the sound waves of the whistles and clicks resembled a primitive form of hieroglyphics.

Schnöller realizes how far-fetched all of this might seem, and he's determined not to go down what he calls a "New Age bullshit path." All data he collects will be analyzed by established researchers in the field; all papers DareWin publishes will be peer reviewed first. "This will be real science," he declares.

SCHNÖLLER HAS GOOD REASON TO be defensive. He's following in a long line of researchers who have lost their minds or, at a minimum, their reputations by attempting to crack the cetacean language code. And no scientist is more representative of this crew than Dr. John C. Lilly, a neurophysiologist who began his career at the National Institute of Mental Health.

In 1958, during one of his first dolphin experiments, Lilly recorded a click-and-whistle conversation between dolphins and played it back at a slower rate. When he adjusted the frequency and speed of these dolphin sounds in water to match human speech in air, he found the ratio worked out to 4.5:1. This was a remarkable discovery. Sound travels 4.5 times faster in water than in air. The frequency of communication the dolphins were using, if modified to the density of water, Lilly wrote, matched the exact frequency of human speech in air. When he played the dolphin sounds at this slower speed, they sounded startlingly similar to human speech. Lilly concluded that dolphins were speaking a language similar to ours, but at a much faster speed, one far too rapid for us to comprehend. He announced his discoveries at an American Psychiatric Association meeting in San Francisco later that year and made international headlines.

By the early 1960s, Lilly had built a sprawling two-story compound that featured a thirty-thousand-gallon saltwater pool and a multiroom office/laboratory complex along the shore of St. Thomas in the U.S. Virgin Islands. The sole purpose of this complex, which he named the Communications Research Institute, or CRI, was deciphering dolphin language.

In 1961, he joined renowned scientist Carl Sagan and Nobel

Prize–winning chemist Melvin Calvin, among other esteemed as-trophysicists and intellectuals, in a semisecret group called the Order of the Dolphin. The purpose of the order was to commu-nicate with extraterrestrials; its first goal was to crack the dol-phin language code. Members wore bottle-nosed-dolphin badges. They traded coded messages. Then they began experimenting. Sagan visited Lilly at CRI several times to help design lab tests, which Lilly started running.

In one experiment, Lilly took two dolphins and placed them in separate pools located at opposite ends of the laboratory build-ing. Inside each pool was a hydrophone and speaker that would transmit sound between the two rooms – an intercom of sorts. Lilly would leave the dolphins alone in their rooms and monitor their behavior in his sealed office. Whenever he opened the lines between the rooms, the dolphins immediately started emitting whistles and clicks. The pool in each laboratory was just a couple of feet wide and a few feet longer than the dolphin's body; the an-imals couldn't be using these sounds for echolocation. They were talking to each other.

Lilly discovered that each of the dolphins' two phonic lips can operate independently of the other; one lip can whistle while the other clicks, and vice versa. During the experiments, sometimes one dolphin would click while the other whistled; other times, a single dolphin would click *and* whistle while the other remained silent. To the untrained ear, these vocalizations sounded cacoph-onous, but when Lilly studied recordings of them, he noticed that the exchanges were always consistent, in that the dolphins would never send clicks or whistles while the other dolphin was send-ing clicks or whistles. In other words, they never talked over each other.

The dolphins, Lilly deduced, could hold two separate, simul-taneous conversations with two separate modes of communica-tion, clicks and whistles – the equivalent of a human talking on the phone while chatting online.

When Lilly shut off the telephone, the conversation imme-diately ended, but the dolphins would repeat the same whistles over and over, as if to say, *Hello? Hello?* The results of the experiment were published in *Science*.*

Lilly was convinced dolphins were communicating in a language that was far faster, more efficient, and sophisticated than human speech. But he still had no idea how to translate the whistles and clicks into English. He continued the intercom experiments, regularly publishing the results in *Science* and other peer-reviewed journals.

In the mid-1960s, however, Lilly seemed to go astray. Against the wishes of Carl Sagan and the other Order of the Dolphin members, Lilly began a battery of wild and often abusive experiments, hoping for a breakthrough. He injected LSD into some animals and monitored their actions. His thinking was that the psychedelic might spur them into suddenly speaking English. (It only made them extremely friendly and vocal.) Then he decided that, since dolphins were so much smarter than humans, it might be easier just to teach them to speak English. While dolphins didn't have vocal cords, they had blowholes that Lilly believed could flex enough for them to form human sounds.

In 1965, Lilly began the first English-immersion workshop for dolphins.

Leading the workshop was CRI research assistant Margaret Howe, who agreed to spend ten weeks doing a wet lab with a rambunctious male dolphin named Peter. By day, Howe would give English lessons to Peter, feed him, and interact with him. By night, she would roll up in plastic sheets on a floating bed in the

---

* In 1963, Lilly's findings were replicated by researchers at a laboratory in Point Mugu, California. The test dolphins, named Doris and Dash, were placed in separate soundproof laboratories and connected by an intercom similar to Lilly's. The researchers recorded the dolphins separately, then disconnected the line. They played back a recording of everything Doris had said in the conversation to Dash. Dash responded to Doris just as he had done before but then stopped thirty-two minutes into the tape. The next day, the researchers played back the recording of Doris to Dash again. Dash stopped talking at the same time. They repeated the experiment yet again, with the same results. The scientists plotted the whistles and click trains and identified a particular whistle signature, which they believed was used as a warning word, meaning something like "Shut up. Someone is listening!" But they could never conclusively determine what it was.

middle of a pool and sleep there while Peter bobbed in the water nearby.

The experiment was a disaster. Howe had trouble sleeping; the constant humidity in the laboratory sapped her energy; she started getting skin infections. In the first three weeks, Peter became sexually aggressive. When Howe swam in the pool, Peter would push her into a corner and thrust his erect penis against her legs. By the fifth week, Peter had grown so obsessed that he had trouble focusing on his English lessons. Howe finally submitted to his sexual advances.

"I found that by taking his penis in my hand and letting him jam himself against me, he would reach some sort of orgasm, mouth open, eyes closed, body shaking," she later reported. "Then his penis would relax and withdraw. He would repeat this maybe two or three times, and then his erection would stop and he seemed satisfied."

On one level, it worked. Peter took a renewed interest in English lessons. His inflection and pitch improved, and he could clearly pronounce simple words like *ball, hello,* and *hi.* He started talking in "humanoid" language when he was alone. When Howe talked on the telephone to people outside the wet lab, Peter became jealous and would speak English words louder to get her attention. When Peter approached her with an erection, she recalled, "I feel extremely flattered at Peter's patience with me in all this . . . and am delighted to be so obviously 'wooed' by this dolphin."

IN THE END, HOWE BELIEVED Peter's English had vastly improved and she was certain that through further instruction, he could develop his vocabulary and perhaps hold a conversation. The results from the English-language-immersion workshop were scientifically inconclusive, but for Lilly they served as proof that humans would be speaking with dolphins within a decade.

Lilly wrote that dolphins would soon be phoning into meetings of the United Nations. They would be starring in television shows. They'd produce underwater ballets, sing pop hits on the radio, and work in underwater industries. But as the years wore on at CRI and the promise of interspecies communication failed to advance much, Lilly grew despondent and depressed. He became ashamed of his work at CRI, where he'd run, as he later put it, "a concentration camp" for dolphins. In 1968, three of the dolphins at CRI died. Lilly believed they'd committed suicide by forcing themselves to stop breathing. He shut down CRI and let the lab's other dolphins go free.

Lilly left St. Thomas and spent most of the next five years in a sensory-deprivation tank high on ketamine, a powerful animal tranquilizer. In 1972, President Richard Nixon signed into law the Marine Mammal Protection Act. It banned the killing, capture, harassment, import, export, or sale of any marine mammals within the United States. The law protected dolphins and whales from slaughter, and it also prohibited scientists from studying wild dolphins anywhere in U.S. waters.

"LILLY BASICALLY RUINED THE FIELD for the next thirty years," says Stan Kuczaj, an experimental psychologist who runs the Marine Mammal Behavior and Cognition Lab at the University of Southern Mississippi. "He did some really great research at the beginning. Those reports in *Science* were really solid," says Kuczaj. "But he just fell off the deep end."

It's around 6:30 a.m. on my fourth day in Réunion. I'm standing with Kuczaj in a weed-filled lot at the La Possession marina. Behind us are two rusty shipping containers that will serve as Dare-Win's field office and the conference center for this week's events. The meeting area is located in the shade of a plastic tarp strung up between the two containers. Seating comes in the form of a few dozen mismatched patio chairs and two wooden stumps. At the center of the seating area is an old door covered in a plastic trash bag and supported by milk crates. This will serve as the conference demonstration table. If we're hungry, there's a box of instant noodles inside one container and a microwave to heat them up.

Kuczaj, who looks a bit like Tom Petty, has been studying dolphin behavior and communication for twenty-five years and is considered one of the world's top scientists in the field. He came to Réunion in part because studying wild dolphins in the United States is prohibited. But what really interested him was Schnöller's archive of wild dolphin and sperm whale footage, which he called "exceptional and extraordinary."

Every morning, Kuczaj and the rest of our group will meet at the DareWin conference center to drink coffee and eat croissants before taking a ride out along Réunion's coast to search for dolphins and whales. If we spot any, we'll stop, get in the water with an assortment of video cameras and audio recording devices, and document as much of the encounter as we can. Around noon, we'll motor back to La Possession and reconvene to share footage on Schnöller's laptop. Every evening, a scientist will present new research to the panel and contribute to Schnöller's action plan for cracking the cetacean language code in the next few years.

Although Kuczaj is very skeptical that humans will ever have a conversation with cetaceans, he's certain that, if it does happen, it won't be through our language but through *theirs*.

He mentions interspecies research done with Koko, a gorilla born in the San Francisco Zoo in 1971 who learned to understand a thousand signs in American Sign Language; and Kanzi, a bonobo who, during the 1980s and 1990s, learned more than three thousand English words.

"Koko and Kanzi might have heard us," he says, "but they only comprehended us in some very limited way." The problem, Kuczaj says, is that researchers have no idea if gorillas or chimpanzees have the capacity to communicate by sounds with one another, let alone with other species. If they don't, then researchers have been trying to teach Koko and Kanzi not just English, but verbal communication — a huge leap.

Dolphins, however, most likely already share a very rich vocal communication. If humans are ever going to talk to the animals, he says, DareWin's approach of trying to crack the whistle-and-click language they already use seems like a good place for us to start.

• • •

TODAY FOLLOWS THE SAME PATTERN as the last four days of Schnöller's conference. We wake up before dawn, climb into motorboats, and search the coast for six or seven hours in a vain quest for whales or dolphins. Finding none, we head back to the marina, eat a late lunch, attend an afternoon conference, then an evening conference, drive back to our rooms, sleep for maybe five hours, then do it all over again.

By midweek, I'm beginning to dread Schnöller's morning knock on my door. Being overworked in any environment is unpleasant; it feels somehow criminal on a tropical island. Kuczaj and the rest of the team were hoping to get a few days to relax and see Réunion—but that's not going to happen on Schnöller's watch. There's always too much work to do and too little time to do it.

And so, on the fifth day of the conference, I awake yet again to a knock at 5:20 a.m.; stumble around in the dark until I find my bathing suit; throw a water bottle, sunscreen, and a notepad in a backpack; and rush out to my rental car before Schnöller starts honking and threatening to leave me behind.

On this day, our luck changes. At around eleven o'clock, we're a few miles off the coast of La Possession. Suddenly, Schnöller stops the boat. "Dolphins," he announces. "Grab your things and get ready."

Swimming with dolphins requires patience and persistence. Schnöller told me he sees them only 1 percent of the time he's out looking; he gets to swim with them only about 1 percent of that time. Those numbers are probably exaggerations, but I get the gist of what he's saying: This is hard work, with very few rewards.

"Dolphins must choose to come to you," he says over the rumble of the outboard motors. "You can never go to them." Chasing them might allow us a quick peek from a distance, but if you get in the water, they'll be spooked and will almost always dive deep. Approaching them very slowly at a 45-degree angle allows them time to observe you and decide whether to interact.

We're about a thousand feet away from the pod when Schnöller instructs me to get a mask and prepare to jump. Kuczaj will be my partner.

"Okay, we go," Schnöller says softly. He has Vanessa, a research assistant from Paris who's joined us for the week, take the steering wheel. Schnöller grabs his camera and then points at me. "You get in behind me, yes?" he says. I nod. The dolphins are in hot pursuit of a school of fish swimming parallel to our boat. Schnöller dips quietly into the water with his camera and kicks away.

Dolphins can be vicious hunters. Schnöller once watched from a boat as a pod attacked a school of five-foot-long tuna. The dolphins swam in circles to gain speed, then shot their pointy noses into the sides of the enormous fish like spear tips. The water soon ran brown and red with blood. (Schnöller got in the water anyway and shot some amazing footage.)

There are no tuna in the water now, at least none we can see. Kuczaj and I pull on our fins and masks and dive in, swimming in the front of the pod in order to intercept its course.

Dolphins get nervous with big groups of people in the water. Having two small groups — Schnöller and a local freediver, Kuczaj and me — enables them to choose whom they'd like to swim with. If they don't like either of us and move on, we must not pursue them. They've chosen not to interact, and we have to respect that choice.

I look up and see that dolphins are now just two hundred feet away. They are swimming directly toward us.

Schnöller motions for us to stop. It's important that we stay as calm and still as possible, so the dolphins don't feel threatened.

While Kuczaj and I float on the surface, Schnöller, about fifty feet ahead, freedives twenty feet down with his camera, ready to capture the encounter. So far, we can't see anything — the visibility is only a hundred feet today, poor by Réunion's standards. But we can certainly hear the dolphins; their clicks sound like a hundred typists pecking away on old Underwoods. This cacophony strikes me as urban and dissonant, something I'd never imagine could come from the natural world.

As I sit there floating stomach-down with my head in the water, I realize that while I can't see the dolphins from this distance, they are watching me. Each of the clicks I hear is bouncing off my body and back to the dolphins, developing in their brains like a thousand little snapshots.

Today's encounter lasts a few minutes, then the clicks fade; the slick backs of the dolphins recede toward the horizon, and they're gone. We turn and kick back to the boat.

"They did not want to play today," says Schnöller. "They must be hungry. Tomorrow—we'll get them tomorrow." Schnöller starts the motor and heads back to the marina. He doesn't seem at all disappointed. And neither am I; I've finally felt echolocation in action.

IT'S SUNDAY, MY LAST DAY IN RÉUNION, and Schnöller and I are sitting around a large wooden table on the front deck of a rented studio apartment in the back of the family's house. Up a staircase to our left is an empty swimming pool, its concrete floor covered with wet leaves, dirt, and puddles of oily water, dark brown like coffee. A robotic pool cleaner, tangled and broken, lies mockingly in the deep end. Above the pool, through an unwashed corner window of the house, sunlight illuminates a room piled high with sofas, board games, clothes, and other old junk. Schnöller's disheveled office is up a cracked staircase at the back of the house. The whole scene looks like a stage set for a tropical-themed *Grey Gardens*. When Schnöller and I talk, as we have most nights after the conference, we meet in my room to avoid having to tiptoe through the rubble.

When I arrived in Réunion ten days ago, Schnöller mentioned that he might be on the cusp of a "big discovery" in cetacean click-and-whistle language but wouldn't tell me what it was. I hounded him all week, but between the DareWin conference, helping out at Planet Nature, and taking care of three kids, he hasn't had the time. Two hours before I leave for the airport, he declares himself ready to give me specifics.

"This is very insane," Schnöller says, repeating his favorite word. "And it's very hard to understand at first, so you must be patient."

Schnöller tells me that scientists know that dolphins use first-name signature whistles to identify themselves in pods, and that they use pod-specific dialects to identify where they've come from and who they're traveling with. Whether dolphins and whales use echolocation clicks as some form of sophisticated language is still a mystery. This is one of the things DareWin hopes to figure out.

But beyond aural communication, Schnöller believes these cetaceans also share a visual language, something called holographic communication. This nonverbal form of communication allows cetaceans to share fully rendered three-dimensional images with other cetaceans, the same way you might snap a photograph on your smartphone and send it to a friend. Schnöller believes cetaceans can share what they're thinking and seeing with one another without ever opening their ears, or their eyes.

Holographic communication sounds far-fetched, but it's not that much of a stretch compared to what cetaceans have already been doing for some fifty million years. Schnöller believes that, since cetaceans can already construct sonographic images from sounds, they might be able to replicate these images and send them elsewhere.

This concept isn't new. In 1974, a Russian scientist, V. A. Kozak, proposed that sperm whales used a video-acoustic system that enabled them to translate the echolocation information into images. Lilly believed sperm whales use sonographic images to communicate, but neither he nor Kozak ever tested the hypothesis.

In the year following the Réunion conference, DareWin researchers plan to conduct the first scientific tests of holographic communication using wild dolphins and sperm whales.

"This is how it will work," Schnöller says, pulling a chair up to the kitchen table. He takes a pen from his pocket, flips my

notebook open to a blank page, and starts drawing dolphin figures surrounded by what look like billows of smoke. That smoke
represents sound, he says; the circle beneath each dolphin's head
represents the jaw.

Sound doesn't travel in a straight line, the way it looks on a
spectrogram, but instead expands in three dimensions, like a
mist. Ears process sound from only two channels; cetaceans have
the equivalent of thousands of channels that can collect this mist
from all directions. "The jaw is just like a sonogram," Schnöller
says. "Only it's in very high definition."

For humans to perceive sonographic images through echolocation isn't easy. Scientists would need to construct an artificial
jaw filled with thousands of little microphones to mimic the tiny
receptors, then build a computer capable of processing all the
data collected. Few scientists have the interest or funds for such
a venture.

Schnöller and Markus Fix, DareWin's lead engineer, are instead building a low-fi version of the cetacean jaw out of a panel
of ten hydrophones wired in series. "The image will be very low
quality—like a ten-pixel image on a computer," says Schnöller.
"But it might be enough to give us an idea." Schnöller plans to
record the "sonar images"—basically, the echo of dolphin and
sperm whale clicks—then process this sound through software,
and play it out of a panel of thirty-nine speakers to gauge the dolphins' reaction. "We have to be careful, you know," he says. "We
don't want to send a picture that is negative or violent."

Through this primitive visual exchange, Schnöller hopes to
make the initial steps of contact with these animals. We'll see
how they view the world, then we'll send pictures of our world
back to them, the way two ancient travelers from different lands
might have drawn symbols in the sand.

BY 6:00 P.M., THE SHADOWS OF THE bamboo bunch outside the
rental have grown long, and the sun is low and lazy-looking on
the horizon. The mosquitoes are out. It's time for me to pack for
my thirty-six-hour flight home.

Before I leave, Schnöller mentions that he's planning a Dare-Win expedition to record sperm whale clicks with some new equipment that he hopes will help with his holographic research. The team will leave in about four months.

I can come under one condition: I must learn to freedive.

# −2,500

ERIC PINON IS SHORT AND narrow, with sleepy eyes, thinning hair, and a meticulously trimmed Fu Manchu mustache. On land, he walks softly, speaks with a slight stutter, and his demeanor verges on meek. But get Pinon in water and he will destroy you. He once speared an eighty-two-pound giant kingfish — stabbed it in the gut at six stories deep — chased it into a cave, rammed his hands inside its gills, and rode it to the surface like a bucking bronco. He can hold his breath for more than five minutes and dive to depths below a hundred and fifty feet.

But Pinon didn't drive three hundred miles from his home in Miami, Florida, to a concrete-block classroom in Tampa to teach us how to kill things in the ocean. He wants to show us how to survive in it.

Thirty years ago, Pinon died. He was freediving with some friends near a pier in the Caribbean and wanted to impress the group with an extra-long breath-hold. So he dove down ten feet, grabbed a pylon, closed his eyes, and tried to stay there as long as he could. Minutes passed. Somewhere along the way, he blacked out. Eventually, his body floated to the surface; he unconsciously exhaled all the air from his lungs, and then he inhaled water and sank like a stone back down to the seafloor.

His friends were impressed when they saw him surface and then sink down again; they thought it was all part of a performance. A few more minutes passed before they realized something was very wrong. They dove down and retrieved Pinon, then dragged him to the beach. His heart had stopped; there were no signs of life. An off-duty paramedic administered CPR. Pinon's heart started beating but soon stopped again. Fifteen minutes later, an emergency helicopter arrived and airlifted him to a hospital, where he spent eight days in a coma, then three weeks recovering. Pinon suffered permanent brain damage that he says sometimes makes it hard for him to remember things and put words together. He doesn't want that to happen to me and other freedivers.

For the past three years, on weekends away from his job managing a fish-feed company, Pinon has traveled around Florida teaching beginning freediving and safety courses through Performance Freediving International, a freediving school based in Canada. This weekend in Tampa, PFI has rented out a one-story stucco building that looks like it once housed a fast-food restaurant.

My classmates sit in a hodgepodge of patio chairs arranged around four plastic picnic tables. There's Ben, a stocky young guy whose gold necklace peeks out of a torn T-shirt; Josh, Ben's soft-elbowed buddy, who sports rainbow-lens sunglasses; Lauren, a tanned southern belle; and Mohammad, a Qatari student with shaggy black hair and an enormous chrome dress watch. Other than Pinon and me, nobody here is older than twenty-three.

In a few hours, Pinon will teach us how to hold our breath underwater for at least one and a half minutes in a swimming pool right outside the classroom. Tomorrow, we'll travel north to a freshwater swimming hole and learn how to hold our breath while diving as deep as sixty-six feet.

This morning, however, is about safety. Specifically, Pinon will teach us how to stay alive should we ever find ourselves caught in a pink cloud—a hallucination freedivers experience right before they black out.

"The pink cloud is harmless, but you are unconscious," says the forty-four-year-old Pinon, who was born in Toulouse, France, and still has a strong accent. "If you get to the surface and breathe, you are fine. If you don't, then . . ." He pauses. "Then it is not good." Pinon means that we'll die.

He explains that, while we're diving tomorrow, he can take us down to any depth we want; he just can't promise to bring us back up. Each of us is responsible for knowing his or her limits. The underwater breath-hold training today and tomorrow will give us a feel for those limits. Should we fail in our responsibility, exceed our limits, and drift off forever into the pink cloud, the six pages of release forms we've each signed will ensure that our loved ones can't charge Pinon or Performance Freediving International with third-degree murder. Pinon double-checks that he has our forms. Then he clears his throat, strokes his mustache, and starts the lesson.

WHILE MY RESEARCH IN THE OCEAN will soon take me down to 2,500 feet, my personal experience is lagging far behind, at just a dozen feet. After many months of observation, training, and envy, I am still stuck wading and waiting at the surface.

I've watched from the deck of a boat as competitive freedivers plummeted three hundred feet and listened to them describe the full power of the Master Switch, but I've yet to feel the full course of these amphibious reflexes myself. I've visited the ama, hoping to receive some ancient, secret freediving advice, only to be mocked for my ignorance. I've heard Fred Buyle talk for hours about the magnetic connection he feels with sharks, but I still haven't seen a shark in the ocean, much less swum with them. I've spent weeks with Fabrice Schnöller and heard him describe the transcendent feeling of communing with dolphins and whales, but I haven't seen these animals either.

What's kept me back is one simple fact: I can't freedive. This ability may be open to everyone, but the price of admission is high: extreme ear pain, claustrophobia, and uncontrollable convulsions. Now, however, Schnöller has offered to let me join him in a dive with sperm whales, an opportunity I can't pass up, and one that requires me to go deep.

Performance Freediving International, considered the best school of its kind in the world, has trained six freediving world-record holders and more than six thousand recreational divers, including Woody Harrelson and Tiger Woods. The entry-level course, called Freediver, teaches students basic safety, depth, and breath-hold techniques. Though I practiced some of these with Hanli Prinsloo in Greece, it's best to just start fresh.

Inside our classroom, Pinon walks over to a laptop and pulls up a video. PFI emphasizes how dangerous freediving is and trains you, right from the start, to deal with potentially deadly situations. He starts with a highlight reel of accidents — the underwater equivalent of those *Red Asphalt* shockumentaries shown in driver's ed.

"This is called a samba," Pinon says. "It's like a dance." Blazing rock music pours from the speakers, followed by clips of divers in the throes of seizures. This pre-blackout state occurs at the surface, when a diver's brain is so deprived of oxygen that it begins sending random electrical signals to the muscles.

"Some people look drunk, some look happy, some very sad," says Pinon. "You see their faces" — he points to the blissed-out face of one diver on the screen — "and they look like they are emotional, in a wonderful dream." Sambas are harmless, Pinon says, as long as divers don't start inhaling water or black out.

After resurfacing, a diver might breathe, start talking, and appear totally normal. But moments later, while air is traveling past the lips, down the trachea, and into the lungs and bloodstream — a process that can take several seconds — he can suddenly lapse into a samba. Should we ever encounter a freediver in a samba state, we must approach gently and hold his mouth above the surface for thirty seconds. Pinon says this is one of many reasons why freedivers should never dive alone, and why we must always watch our diving partners for a full half minute or longer after they surface. He underscores this point repeatedly.

The next video shows divers who have passed the samba stage and lost consciousness. While holding your breath, "you can black out anywhere," Pinon says. "In a deep ocean, in a shallow lake, in a bathtub — anywhere." He says that 90 percent of sambas and blackouts happen at the surface; another 9 percent happen within about fifteen feet of it, what freedivers refer to as the danger zone, the area in the water where the greatest shift in pressure occurs. Freedivers very rarely black out on the seafloor. They black out at the surface, then sink back down and drown, like he did.

The first step in saving a blacked-out diver is yelling "Breathe!" in his ears and calling his name. In the blacked-out state, vision and physical sensation disappear, but hearing remains, and it's often heightened. Yelling, Pinon says, activates parts of the brain that have not yet shut down. This jolt can override the body's reflex to close the throat so fresh air can enter the lungs.

If yelling doesn't work, we have to remove the diver's mask, tap his face, and start blowing on his eyes. The technique frequently revives blacked-out freedivers; often, they'll come to and begin gasping for air.

Now, if tapping, yelling, and blowing all fail to wake the diver, Pinon says, "things get more serious." We need to open the throat and force air into the lungs.

One way the body prevents drowning is by closing the larynx when it comes in contact with water. We're all born with this reflex. When a newborn is put in water, his larynx automatically closes; the baby will open his eyes and instinctively begin swimming underwater.

During blackout, the closed larynx will keep water out of the lungs (a good thing), but it will also keep fresh air out of the lungs (bad). Many drownings in water are known as dry drownings, meaning they result from the larynx closing, not from water getting into the lungs.

Pinon shows us how to open a diver's mouth with our fingers and breathe two quick puffs. The first opens the larynx; the second delivers air into the lungs and stimulates the body to begin breathing again. In almost all cases, Pinon assures us, this will bring a blacked-out diver back to consciousness.

While such rescues can be frightening and stressful, actually having a blackout is anything but. "All the pain just goes away," says Pinon, smiling.

It starts with visual disorientation and light hallucinations, then the fingers, toes, hands, and feet start to tingle. You lose muscle control. These symptoms progress until you enter a state of spacy euphoria accompanied by wildly colorful dreams, the aforementioned pink cloud. Blacked-out divers have reported out-of-body experiences: One competitive diver in Greece told me he saw into the future. (What exactly he saw, he wouldn't say.)

However mind-expanding blackouts can be, it's obviously best to avoid them. Extended blackouts are occasionally fatal, and when they're not, they can cause brain damage, paralysis, cardiac arrest, and strokes.

Every second you hold your breath, oxygen begins to drop. If the oxygen in the brain drops below a certain level, you black out. A person can stay in this blacked-out state safely for about two minutes until brain oxygen gets so low that you enter what's called an anoxic state. Anoxia will trigger the body to initiate a last-ditch effort to breathe, called a terminal gasp. If there is no oxygen available at this time (for instance, if you're underwater), brain damage begins to occur and you'll eventually die.

The key to avoiding a blackout, and the resulting brain damage, is getting to the surface as soon as you start feeling spacy, lose muscle control, or experience hallucinations — a hard thing to do if you've miscalculated your dive ability and your muscles start convulsing at two hundred feet down. "This is another reason to always dive within your limits," Pinon says emphatically.

He dismisses the class for lunch, advising us to eat light, preferably something vegan and caffeine-free. Dairy, he says, can plug the sinuses and make it hard to equalize at depth. Caffeine will raise the heart rate and speed up metabolism, causing the body to suck up more oxygen and shortening dive times. After lunch, we'll all get in the pool and start testing our limits.

OF ALL THE DISCIPLINES IN freediving, static apnea, a timed breath-hold that usually takes place in a pool, is the strangest. It's boring to watch, painful to do, and tedious to train for. And yet there is no other activity that will better prepare a freediver to handle the mental and physical stresses of deep diving.

In 2001, the world record for static apnea, held by a Czech named Martin Štěpánek, was just over eight minutes. In 2009, Stéphane Mifsud, a French diver, increased the record by 27 percent, to eleven minutes and thirty-nine seconds.* As of 2013, two divers have held their breath for more than ten minutes: Mifsud and Tom Siestas, of Germany. If static divers continue at their current rate, they'll break the fifteen-minute mark mentioned in historical accounts of pearl and sponge divers by around 2017.

Static apnea has its own set of fringe disciplines: breath-holds in shark tanks, under ice, in plastic bubbles. An increasingly popular variation is static with pure oxygen, which follows the same rules as regular static apnea except that divers can huff pure oxygen a half an hour before going under. Doing this supersaturates the blood with oxygen, allowing the brain and other organs to function significantly longer than they could if the diver had inhaled natural air (which contains only about 20 percent oxygen). David Blaine, the American magician and stuntman, trained with PFI and in 2008 broke the static-apnea-with-oxygen record with a breath-hold of seventeen minutes and four seconds that he performed live on *Oprah*. Five months later, Siestas broke Blaine's record, and Siestas now holds the all-time record: an astounding time of twenty-two minutes and twenty-two seconds.

---

* In January 2012, Branko Petrović of Serbia broke Mifsud's record with a twelve-minute, eleven-second static breath-hold. However, this record has not been certified by AIDA because Petrović had no international judge present during the dive.

-2,500

We won't try anything like that in our course. To be officially certified as a freediver, each of us needs to do a static breath-hold of at least one minute and thirty seconds. Physically, it's not much — any human in decent health is capable of hitting that mark. But mentally, it can be a challenge. There is nothing intuitive or natural about keeping your face underwater until your brain begins hallucinating and your muscles convulse. But this, I am told, is all part of going deep.

BY 1:30 P.M., WE'VE SLIPPED on wetsuits and regrouped in the shallow end of the pool. PFI's intermediate class, which is being held in an adjacent room, enters the pool too. One of two male intermediate instructors stands at the side of the deep end with his shirt off. He has tattoos of fish gills running up both sides of his rib cage.

Mohammad, the quiet Qatari with the chrome watch who was seated next to me in class, agrees to act as my monitor. He'll periodically check that I'm still conscious and will keep my body from drifting.

The sensation of spinning is common during long breath-holds, because your body loses awareness of its own boundaries. This is a hallucination, says Pinon, but nothing to worry about. While placing a hand on the back does not keep static divers from accidentally losing consciousness, it reassures them in their hallucinatory state that they aren't suddenly sinking, floating off, or flying away.

Pinon gives a one-minute warning. I slip my mask on and start breathing a little deeper. Pinon and Mohammad chant the pre-dive breathing pattern aloud: "Inhale one, hold two, exhale two-three-four-five-six-seven-eight-nine-ten, hold two." On Pinon's command, I take four huge breaths and then sink headfirst into the water.

I can do the one-minute hold with no problem. A few minutes later, the two-minute hold is filled with tedious agony. But, strangely, the three-minute hold goes by in a comfortable haze for me, as though I've crossed some invisible border. I don't pass out, and for a few minutes afterward, I feel lightheaded, dizzy, and very high, like I've just huffed laughing gas. It's great.

Feeling this good would ordinarily be damaging in some way, at a minimum killing off a few thousand brain cells. But according to dozens of studies, extended breath-holds are harmless. Neurological damage occurs when the blood in the brain carries too little oxygen or when blood flow stops completely. These conditions occur only after two minutes in a blacked-out state. In other words, as long as you're conscious or wake up from a blackout within two minutes, there's a very good chance you'll suffer no damage from holding your breath. Water extends your time by shunting blood from the extremities into the brain and organs, allowing them to function with minimal oxygen for much longer than they would on land — triggers of the Master Switch.

Under normal conditions, the human body has a blood-oxygen saturation of around 98 to 100 percent (the higher number being the most oxygen that the blood could possibly contain). Physical stress or sickness can decrease oxygen saturation to about 95 percent. Few healthy people will ever go below this, but during dives, expert divers have registered oxygen-saturation levels as low as 40 percent — an extraordinarily low number. Oxygen saturations below 85 percent generally cause an increased heart rate and impaired vision; 65 percent and below greatly impairs basic brain functions; 55 percent results in unconsciousness. But somehow, expert divers have not only remained conscious with oxygen saturations lower than 40 percent but maintained muscle control and extremely low heart rates, reportedly as low as seven beats per minute.

BACK AT THE POOL, my class is getting ready for the final breath-hold of the day, which will last four minutes, the maximum allowed for this introductory course. During the longer holds, partners have done periodic check-ins by tapping the breath-holders on the shoulder every fifteen seconds. When a diver feels the tap, he has two seconds to extend the forefinger of the submerged left hand, a way of saying, *I'm still here, I'm okay.* If he doesn't respond, his partner will give him one more chance and tap again. If the second tap elicits no response, the diver's partner will lift him from the water, yell at him to breathe, remove his goggles, and blow on his eyes.

The partners begin chanting the warm-up breathing pattern —"Inhale, exhale, hold two-three-four-five-six-seven-eight-nine-ten, hold two, inhale one." The intermediate class at the deep end of the pool joins in the chant. I'm still quite high from the three-minute attempt and feeling spaced-out as I breathe deeper. The chorus of voices echoing off the concrete walls grows louder, reverberating around the enclosed pool area like incantations in an old church. It's hypnotizing. The course is beginning to feel like a baptism, each of us trying to be reborn in a watery world.

Then it's one more breath, and we're underwater again.

A minute passes, then two. Every fifteen seconds, Mohammad taps my shoulder. I extend a finger, bend it down, extend again. During the second minute, I notice sounds in the pool area that I hadn't heard before: a gurgle in the drain, a muffled cough, a splash in the deep end. I hear Mohammad counting somewhere over me, feel his hand on the small of my back, then stop feeling much of anything. I imagine myself traveling in a train through the desert. This scene looks very real. One part of me knows that I'm still in a Tampa swimming pool, but another part seems convinced that I've boarded a faraway train. Both parts are equally strong, like reflections of each other. As my stomach begins to convulse, I push my mind farther into the train side, to open that door wider.

A conductor announces that we'll be disembarking in three minutes. He taps me on the left shoulder and I hand him my ticket with the index finger of my left hand. The blue fabric of the seat is soft, like silk. I stroke it with this finger. The conductor taps me on the shoulder again; I reach in my pocket to hand him the ticket, but the ticket is gone. I motion with my finger for him to wait while I look in my bag. I can't find my bag. The cabin is too dark; the sun is gone. I hear someone nearby splashing water in a sink. The conductor taps me again on the shoulder. I point to the door and ask if I can get off. *You can do this,* he says. *You can do this.*

I come to, my head still in water, and I'm staring through my mask at the pool's white concrete bottom. It feels like someone has filled my lungs with mustard gas. "Three forty-five. Almost there," Mohammad says. I put my hands on the side of the pool to stop myself from sinking, from falling down into what feels like a deep, dark hole.

"Breathe!" says Pinon. I lift my head. "Breathe! Breathe!" says Pinon. The room spins. I try to exhale my lungful of air, but I've lost some muscle control and can't. I push harder to force it out, to get a breath of fresh air in. A puff of air squeaks out, then my throat opens. I exhale completely and take a long inhale. With every intake of breath, my pinhole vision grows larger and larger, like the opening sequence of a James Bond movie. The room is blurry and covered in static for a moment, then everything comes into focus.

The instructor with gills tattooed on his ribs swims over and pats me on the back. "Good job, man," he says. I'm the only one in our class who completed the four-minute breath-hold.

THE NEXT DAY OUR CLASS meets about a hundred miles inland from Tampa in a dirt parking lot outside the town of Ocala. Across the lot, in the shade of candleberry trees, is a dent in the ground that looks as if it's been punched by a giant, angry fist. At the bottom of the hole is a pool of bright green water known as the 40 Fathom Grotto. As the name suggests, it plummets down more than 240 feet.

For the past forty years, emergency rescue workers have used the grotto for advanced scuba training. Before that, the locals used it as a public dump. It's still filled with all manner of junk: rusty motorcycles, satellite dishes, a 1965 Corvette, a few Chevys, an Oldsmobile, innumerable bottles and cans. On one ledge, about forty feet down, is Gnome City, a collection of plaster gnomes and gnome castles placed there by divers; it's set against a limestone wall covered in the fossilized remnants of fifty-million-year-old sand dollars. Even this water hole, fifty miles from the coast, was once part of the ocean.

At ten o'clock, Pinon swims two floats out to the middle of the grotto and connects them with a yellow rope. Our class pulls on wetsuits, masks, snorkels, and fins, and we slip in. In the hazy morning light, the water is a dull sapphire green with poor visibility, maybe twenty feet. The depths below that look black and brooding. We swim out to the floats and clutch the rope, dangling in single file like socks on a clothesline. We'll be here for the next four hours attempting to freedive to sixty-six feet.

"Our first dive will be down to five meters," Pinon says. "This is an easy dive, just to get warmed up." Because the grotto is filled with fresh water, which is less dense than salt water, we'll be about 2.5 percent less buoyant than we would be at sea. This doesn't sound like a lot, but for freedivers, it's a significant difference. We'll sink faster and will have to exert a bit more energy during our ascents.

The human body in its natural form — with little or no clothing — has the ideal density for freediving; no weights are necessary to aid its descent. However, the thick wetsuits we're all wearing throw off this balance, requiring each of us to wear about twelve pounds of weights in fresh water to compensate for the extra float.

The key to a successful deep dive is making oneself as hydrodynamic as possible. Loose clothing, extended limbs, or oversize masks can create drag, which will slow the descent and decrease depth and "down time" — freediver lingo for being underwater. When seals dive deep, they collapse their lungs, extend their spines, and often exhale air to reduce drag and gain depth faster and more easily. Freedivers do the same. "You put your arms to the side, head down, make yourself like a missile," says Pinon.

Sinking is relatively easy, especially after the first ten or so feet; ascending is less so, which is why freediving can be so dangerous. As with mountaineering, you need to know your exact halfway point and have at least 60 percent of your energy and oxygen reserves left to make the return trip.

During the ascents, we'll need to exhale all the air we've been holding at about seven feet below the surface. This allows us to immediately inhale much-needed fresh air at the surface without taking time to exhale, and it also helps protect against shallow-water blackouts. A few seconds could mean the difference between a successful dive and a samba or blackout. In freediving, success (in this case, remaining conscious) is measured not in feet or minutes but in inches and seconds.

As Pinon discusses the diving strategy, I notice a small group of scuba divers on a wooden float set up on the other side of the grotto. They are festooned head to toe with masks, tubes, tanks, vests, belts, and other equipment. They can barely walk on land and can only lumber gracelessly through the water. Their movements are extravagant because they can afford to be. From where I'm floating, it looks awkward and wasteful. But then again, those divers never have to worry about imploding their lungs or blacking out.

Ben dives first. We watch through our masks as he breathes up, submerges, and pulls himself down along the rope until he reaches a weighted plate around fifteen feet. He taps the plate, pulls himself back up, resurfaces, and goes to the end of the line. Lauren, Josh, and Mohammad, one after the other, go next. They all make the dive without much effort. I follow but resurface after hitting just ten feet or so, my head throbbing.

"That's natural," Pinon says. "It takes a little time. Try it again next dive."

I ask Ben how he was able to descend and ascend so quickly. He mentions that he, Josh, and Lauren have been spearfishing for years. He assures me I'll figure it out.

The problem for me, and for most beginners, is equalizing. The optimal rate of descent for a freediver is three feet per second, which requires equalization in sinus cavities (making the ears pop) about once a second, otherwise you'll risk serious in-

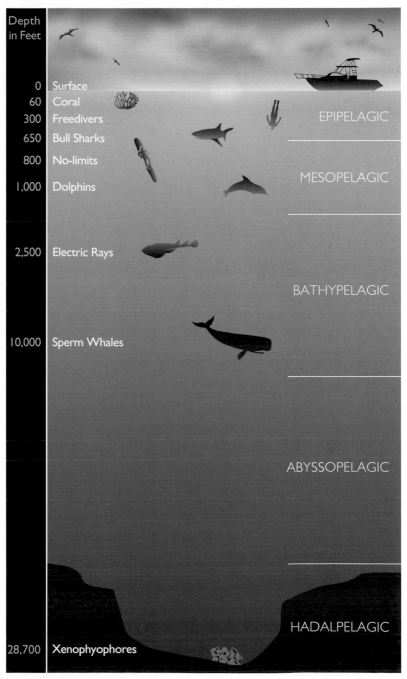

| Depth in Feet | | |
|---|---|---|
| 0 | Surface | |
| 60 | Coral | |
| 300 | Freedivers | EPIPELAGIC |
| 650 | Bull Sharks | |
| 800 | No-limits | |
| 1,000 | Dolphins | MESOPELAGIC |
| 2,500 | Electric Rays | |
| | | BATHYPELAGIC |
| 10,000 | Sperm Whales | |
| | | ABYSSOPELAGIC |
| | | HADALPELAGIC |
| 28,700 | Xenophyophores | |

Illustration by Josh Ceazan

Sperm whales produce clicking sounds that allow them to view their underwater world with echolocation, a form of sonar. DIY researcher Fabrice Schnöller believes these sounds may also contain a kind of coded language. He built this homemade contraption of surround-sound hydrophones and video cameras to better study sperm whale vocalizations and behavior.
*Fred Buyle/Nektos.net*

The eye of the whale: The sperm whale brain is six times the size of ours and in many ways more complex. It's the largest brain ever known to have existed, and most researchers believe it allows sperm whales to engage in sophisticated communication. Schnöller and his team hope to be the first to decode what they're saying. *Fred Buyle/Nektos.net*

"It's like being in outer space": At underwater depths below forty feet, gravity reverses. Instead of floating up, you're pulled down. Hanli Prinsloo, a champion freediver and ocean conservationist, tries out her deep-water high-wire act. *Annelie Pompe/anneliepompe.com*

Prinsloo enters a shiver, or group, of blacktip sharks. When freediving researchers approach sharks "on their terms"—that is, deep below the ocean's surface and without the disruptive gurgle of air bubbles from scuba equipment—the threat of attacks is often negated. The sharks become curious, even docile. *Jean-Marie Ghislain/ghislainjm.com*

"They're gentle giants": Whale sharks are neither whales nor sharks, but the world's largest fish. They can grow more than forty feet in length and weigh around 50,000 pounds. Prinsloo swims face-to-face with one as it feeds on plankton. *Jean-Marie Ghislain/ghislainjm.com*

Dolphins "speak" both first names and last names when they approach other dolphins and, sometimes, human freedivers. They may also be able to send one another sonographic pictures, something researchers call "holographic communication." Schnöller dives down to take a closer listen. *Olivier Borde*

"Jane Goodall didn't study apes from a plane": Of the twenty or so sperm whale scientists who work in the field, none dive with their subjects. Schnöller (center right, with camera) finds this inconceivable. "How do you study sperm whale behavior without seeing them behave, without seeing them communicate?" With his immersive, freediving approach, in five years Schnöller has captured more audio and video footage of sperm whale interactions than anyone before. *Fred Buyle/Nektos.net*

The key to getting close to whales and dolphins, Schnöller says, is to move in calmly and predictably. With a bit of luck, humpback whales (pictured here), normally very shy, get playful. Sometimes, they approach. *Yann Oulia*

"They stayed with us for three hours": A freediver approaches a blue shark in this image captured by Fred Buyle, one of the world's most sought-after underwater photographers. Buyle aims to document the gentle side of what he calls "the most misunderstood animals on the planet." An estimated twenty million blue sharks are killed every year for shark fin soup and fishmeal. *Fred Buyle/Nektos.net*

The human body in its natural form (without weights or wetsuits) is the perfect buoyancy for deep-water diving. We're able to float at the surface and yet can dive down to great depths while exerting little effort. From left, swimmer Peter Marshall, Hanli Prinsloo, and the author. *Jean-Marie Ghislain/ghislainjm.com*

As on the moon's surface, gravity on the seafloor still exists below forty feet but is greatly reduced. Strolling around, doing yoga, even underwater hiking are possible for as long as you can hold your breath. William Winram, a Canadian freediver, takes an amble. *Fred Buyle/Nektos.net and williamwinram.com*

jury to the ear. Each pop must be complete; if it isn't, Pinon instructed us to immediately stop, back up, and try it again.

Pinon lowers the plate to thirty feet, then forty-five. Others easily make these depths, but I can't make it past fifteen feet.

At around two o'clock, it's time for our last dive attempt. The plate is now sixty-six feet down — the lowest depth allowed for beginners. It's invisible from the surface. All we can see going down is a yellow rope disappearing into dark green water.

It's a frightening prospect to dive down into water not knowing where you'll be when the rope ends or when you'll take your next breath. Everything I know about surviving in the ocean tells me this is a bad idea. But I start breathing up anyway and prepare to dive.

Ben leads the group. He inhales one last time, then disappears. Forty-five seconds pass and we see no sign of him. Then, through the haze, he reemerges, pulling himself up the rope. He slowly resurfaces, breathes up, then goes to the back of the line. He made it down to sixty-six feet, seemingly without much effort. Lauren and Josh follow, all making the dive. Mohammad, a first-time freediver, makes it to about fifty feet, a commendable depth.

By the time it's my turn, the pressure is on. I try not to look down at the disappearing rope as I inhale my last breaths. Big breath in, bigger breath out. Repeat.

Pinon pulls himself around the float so that he's right beside me. "You need to make this dive. Say, 'James is going to make this dive,'" he tells me. I nod, inhale, duck my head under the water, and climb down the rope.

With every pull of my right arm, I retract my right hand, pinch my nose, blow air into my ears, and try to equalize. It starts to work. I keep pulling, hand over hand, like Jack and the Beanstalk in reverse, until I feel the pressure of deeper water tightening around me. To make my body more hydrodynamic, I've placed my head down, so that I'm looking horizontally across the water, as if I were walking. Pinon, who is following me on the other side of the rope, stares through his mask. He is watching carefully to make sure I don't exhale, start twitching, or black out.

I stare back and we hold each other's gaze as we both sink. The water around us grows darker, then darker still. A strange sensation grabs at my shoulders. It feels like a large hand is pulling at me. I loosen my grip on the rope and notice that I'm no longer drifting downwards. Every direction is washed in the same pale green fog — I'm trapped in an enormous marble. I wouldn't know which way was up or down if I weren't holding the rope.

Across the rope, Pinon is looking at me and shrugging. He puts his right hand in front of my mask and points down. *He wants me to go deeper,* I think. I shake my head no but he keeps pointing down. I notice that neither of us is holding the rope.

We're just suspended here, two middle-aged men floating upside down, staring at each other, shaking our heads in the shadowy depths of a freshwater former dump in central Florida.

Then it occurs to me that maybe down is really *up,* and that maybe Pinon is signaling for me to get back to the surface. Maybe something is wrong. Is this what it feels like in the pink cloud?

I snap out of it, but now I really want to breathe. A cough right now could sap my body of the oxygen I need to make it back to the surface conscious. This thought fills me with fear. I feel an unyielding urge to return to the surface, to inhale fresh air. I quickly turn my body around on the rope like a baton and begin pulling back up. Pinon follows close behind. With each pull, the water grows slightly brighter and brighter until I can see, about fifteen feet above me, rows of dangling fins between two floats. They look similar to upside-down birds on a telephone wire. I exhale all my air at what feels like seven feet, then resurface.

I LEARN LATER THAT I made it about halfway down the rope, to about thirty feet. Not awful, but not great either. This wasn't the doorway to the deep, but I was edging closer, starting to wipe my feet on the welcome mat. The tug of neutral buoyancy I felt just before I started back up the rope meant I was about ten feet away. For better or worse, the residual fear of being down there remained with me.

And days later, as I'm in the airport on my way home, I'm still shaking with excitement and looking around before I cough.

------------------------

Someday, freediving may help me swim with sperm whales and, perhaps, plunge well past forty feet, but it could never take me down to the edge of the bathypelagic zone, a realm of pure and permanent blackness that extends from 3,300 down to 13,000 feet down. No sunlight has ever touched these deep waters. The pressure is debilitating, ranging from a hundred to four hundred times that of the surface, and water temperatures hover at a frigid 39 degrees. It's hell, without the heat or crowds.

No diver—neither scuba nor freediver—has ever been down past 1,044 feet, which is just one-third of the way to the bathypelagic. Humans can access this world only by using deep-water machines. A remotely operated underwater vehicle (ROV), a phonebooth-size robot covered in lights and video cameras and attached by a cable to a boat, can plunge tens of thousands of feet deep, but it can carry no human cargo. There are about a half a dozen ROVs in the United States, operated by colleges and oceanographic institutions, that can reach the bathypelagic, but the experience of watching video stream from the deck of a boat strikes me as isolating. Nothing could match the experience of actually being down there.

Few submarines or submersibles could make such a journey. Probably the most famous research submersible (a submarine supported by a surface vessel) in the world is *Alvin,* a U.S. Navy vehicle first launched in 1964. During its past five decades of operation, Alvin has made more than 4,600 dives, many well into the bathypelagic. Hitching a ride on it was impossible. According to the media director at Woods Hole Oceanographic Institution, which operates the vessel, no journalist had ever ridden on the sub, and no nonscientist would be welcomed aboard. (I discovered later that the media director either was playing dumb or was misinformed. On rare occasions, *Alvin* has taken journalists aboard, sometimes on deep dives. But there was no point in arguing with her. It turned out that *Alvin* was in dry dock for the next two years getting upgraded.)

My only other option would be to board a private submarine. In the past decade, a cottage industry of specialized sub manufacturers had sprouted up in Florida and California, and for the first time, it had become possible for hobbyists to descend to about thirty-three hundred feet. These submarines, however, are prohibitively expensive, ranging from $1.8 to $80 million, and they can take years to build. Buying one obviously was out of the question. All attempts I made to contact these subs' super-wealthy owners proved fruitless. I never even got a response.

Then a friend told me about a New Jersey man named Karl Stanley who began building a submarine out of plumbing parts in his parents' backyard when he was fifteen. When he was fourteen, Stanley spent six weeks in a mental hospital — diagnosed with "defiance-of-authority syndrome." When he got out, he successfully proceeded building his DIY sub, with no engineering background. Eight years later, in 1997, he had designed and handbuilt a vessel with the lightest displacement hull (the part of a marine vessel that controls buoyancy) in history, and he'd done it for a total of $20,000 — about a hundredth of what it would have cost on a project run by engineers. The sub, which he called a controlled-by-buoyancy underwater glider (CBUG), could carry two people down to 725 feet.

Because taking tourists down seventy stories in a homemade, unlicensed submarine, without insurance, was a liability nightmare, Stanley moved his operation to Roatan, Honduras, where regulations for underwater craft were lax or nonexistent. Stanley's submarine tours were a hit. A few years later, he designed and built a bigger vessel, named *Idabel,* that could carry three passengers down to about three thousand feet.

The bathypelagic, or midnight zone, is defined as any depth in water where no sunlight can penetrate. In the Caribbean Sea off Roatan, that depth is about seventeen hundred feet, well within *Idabel*'s range. I wanted to go as deep as *Idabel* could take me, because I didn't know if I'd ever be able to go that deep again. Although 2,500 feet was still a few hundred feet short of what oceanographers consider the official bathypelagic zone, it was as close as any private citizen of ordinary means could get.

Better yet, Stanley had no disclaimers, no liability waivers for me to sign, no insurance requirements. If something bad happened, all passengers, including Stanley (who piloted every dive), would die. End of story. No submarine left and nobody to sue. All I had to do was wire him $1,600 and pick a date.

The trip down and back, he said, would take about four hours. For the duration of the dive, I'd be curled up in a steel ball the size of a car trunk, and I would have to stay there without stretching, peeing, or losing my mind for the entire expedition.

Stanley has had his share of close calls. He's gotten CBUG stuck in a cave and snagged on a rope below 200 feet. Once, at 1,960 feet while he was carrying a local from Roatan and his pregnant wife inside, a window cracked but fortunately didn't shatter. On other expeditions, washers have slipped out; occasionally a motor seized or just quit. But Stanley has made design fixes every time, and after nearly two thousand dives in both CBUG and *Idabel,* nobody has died; nobody has even gotten hurt.

In his homemade, hand-built, self-designed submarine, Stanley has spent more time in the deep waters between one and two thousand feet than anyone in history.

THE ROATAN INSTITUTE OF DEEPSEA Exploration, the official name of Stanley's submarine-tour business, sits at the outer edge of the island's touristy West End, a crescent bay of white sand and placid blue water facing the Caribbean. The West End is a popular getaway for backpackers, American families on a budget, and day-tripping cruise-ship guests, and it looks the part. There are sand-floored tiki bars serving pink slushy drinks, sunburned men with balloon stomachs and stick legs, and women with bottle-blond hair wearing pink bikinis. Stray dogs on trash heaps scratch hairless backsides. Locals smoke knockoff Marlboros outside a restaurant called Cannibal Café. The sound of Bob Marley's *Legend* blasts from duct-taped boom boxes and mingles with Nokia ringtones that chirp from the pants pockets of bare-chested taxi drivers.

About a half a mile from all the noise, and down an unmarked road canopied by dead palm fronds, is Stanley's house, an old colonial-style clapboard a few dozen feet from the water. A wooden walkway reaches out to a small dock. Painted across the dock's awning are the words *Go Deeper*. In the shade beneath, dangling from a steel cable, is a sub straight out of Beatles' lore: it's bright yellow, with round windows on all sides and a circular stovepipe bump in the back. The name *IDABEL* runs across one side in blue, sans-serif capital letters.

Stanley stands behind *Idabel* holding a chunk of steel machinery he's just pulled from the hull. Tall and slender, he wears reflective glasses, a gray muscle T-shirt, and khaki shorts. A generator coughs exhaust in the background, and gusts of compressed air fart from beneath the sub every few seconds. I'm early, and I've caught Stanley making a few last-minute fixes. He looks annoyed as I approach to shake his hand, holds eye contact for a little too long, then walks back to the sub without a word.

*Idabel* is double-occupancy, meaning that I had to purchase two passenger seats ($800 each). I asked Stan Kuczaj, the dolphin scientist I had met a few months earlier in Réunion, if he'd like to come along. Kuczaj was staying in Roatan for the week to conduct research on captive dolphins at a resort a few miles away. Even though he'd been studying the ocean and its inhabitants for more than thirty-five years, he'd never been in a submarine, and never seen the ocean past around 120 feet down while scuba diving. He was ecstatic to join me.

But after Kuczaj viewed *Idabel*'s cramped conditions, his chattery excitement turned to a sort of silent dread. The sub is thirteen feet long and about six feet wide. Its top portion has nine glass portholes, allowing for a 360-degree view. Stanley stands in that spot and looks out the ports when he pilots the sub. Passengers ride up front, at his feet, in an area he calls the passenger sphere. This section is a scant fifty-four inches in diameter with about three feet of sitting room, or about the width of a La-Z-Boy recliner. In our case, this chair will have to seat two six-foot-two men. In front of this seat is a thirty-inch-wide convex Plexiglas window.

The great advantage of running submarine tours in Roatan is easy access to the Cayman Trench, a deep-water chasm that runs undersea from near Jamaica to the Cayman Islands. At its deepest, the trench plummets down more than 25,000 feet and contains the world's deepest volcanic ridge. In 2010, a group of researchers from England's University of Southampton sent an ROV down there and discovered the world's deepest and hottest hydrothermal vents. (Hydrothermal vents are underwater volcanoes that spew toxic gas more than a half a mile up from the seafloor.) The temperatures near these vents reached 800 degrees, hot enough to melt lead. A return trip in 2012 revealed that the vents, perhaps the most hostile environment on the planet, were home to a host of bizarre animals and microorganisms — species that have been observed nowhere else on Earth.

We won't be going near any hydrothermal vents today, but we will be diving through pitch-black waters to a spot on the seafloor that no human has ever seen. "I discover something new on every dive," Stanley tells us.

IT'S TIME TO BOARD. KUCZAJ volunteers to hop in first. I watch as he squeezes his long torso through a two-foot-wide hole at the top of the sub, like a recoiling Whac-A-Mole. Once inside, he worms feet-first into the observation deck, bending his legs up to his chest as he crams himself into the tiny seat. He gives a thumbs-up through the front window and shakes his head, laughing. "I don't think you're going to fit in here!" he yells. "I'm serious. You're not going to fit in here."

I prove him wrong, managing to wiggle in next to him. I'm sitting not so much beside Kuczaj as *on top* of him. Because the curved walls make it impossible to lean back, our spines must hunch forward like parentheses. The head clearance is so low we have to crane our necks downward, turtle-like, to avoid scraping our scalps.

Before we change our minds about taking this journey, Stanley crawls in behind us. *Idabel*'s electric motors rumble to life. The cable suspending the *Idabel* unwinds and lowers us into the water. We disengage from the cable. Stanley shuts the hatch, and we start heading north, half submerged, to the Cayman Trench.

In the passenger sphere, Kuczaj and I gaze out the window at two very different worlds: brassy sunshine above and silvery water below. Under direct sunlight, the bubble-shaped front window works like a magnifying glass to superheat the interior. Before we leave the bay, the temperature inside *Idabel* has reached 98 degrees. Kuczaj is sweating through his T-shirt; he looks angry and anxious. Almost immediately, he develops a nervous tic, repeatedly pressing the shutter button of his camera. We take off our shoes to cool our sweating feet.

*Idabel* sinks lower; the light and life from the surface begin to fade.

"There is going to be some rocking and tipping here," Stanley says. With a jolt, he tilts *Idabel* about 45 degrees and hovers for a moment so that we're looking directly into the maw of the Cayman Trench.

"Here we go," Stanley announces. We begin the descent.

At the top of the window, toward the surface, bands of blue stretch out on the horizon and grow darker and darker, like a Rothko painting. The gradients of color aren't a trick of light, or a mirage — they're the spectrum of sunlight getting slowly swallowed up by water molecules.

At the ocean's surface, the sun's energy penetrates easily through the water. Deeper down, that energy fades until, at depths of around three thousand feet, there is no light. Longer-wavelength colors, like red and orange, are easiest for water molecules to absorb, and so they drop out first. The color red becomes invisible to the human eye at around fifty feet down; yellows disappear at around a hundred and fifty feet; greens at two hundred feet, and so on, ultimately leaving only stronger, shorter-wave colors like blue and purple.

The blue ocean water (and sky) we see from the surface has nothing to do with the color of water or air—both, of course, are colorless. Tropical water appears intensely bluish-purple because the visibility extends for hundreds of feet, allowing you to peer into the depths where only blue and purple light can penetrate.

Blue fish are exceedingly rare in the ocean, because they would be highly visible until they reached the lightless waters of the bathypelagic. Meanwhile, red fish are fairly common because red is the best camouflage in deep water. A fish like a red snapper looks red at the surface, but as it descends, the redness appears to fade away until, at around a hundred feet, it becomes virtually invisible to its prey and predators. This is why snapper spend almost all their time at between fifty and two hundred feet.

STANLEY TILTS THE *IDABEL* sharply so that we're pointed almost straight down. We glide gracefully to the ocean floor, like a hot-air balloon in reverse. The depth gauge reads −300 feet. Stanley still hasn't turned on any of *Idabel*'s interior lights, and the monochrome palette washes over us. Our clothes, skin, notepads, and the world outside all appear to be the same bluish color.

A few minutes later we pass eight hundred feet. We've entered the mesopelagic. At these depths, 99 percent of sunlight has been absorbed by water. No plants can survive, and from here on down, the entire ocean is animal and mineral. *Idabel*'s frame begins to fizz and creak under the pressure, which has reached more than three hundred and fifty pounds per square inch. Stanley has calibrated the cabin pressure to match the 15 psi of sea level, but I'm not sure it's working. It feels as if the pressure keeps mounting the deeper we go. Every thirty seconds or so, we have to equalize our sinuses so that we don't blow our eardrums out. Kuczaj is hunched over on his knees and looks like he's going to be sick. Then I'm hit by a wave of nausea and paranoia.

"You okay?" I ask Kuczaj.

"Heavy," he says. "Heavy breaths."

ASTRONAUTS LEARNED TO DEAL with the psychological and physical trauma of space travel by focusing on specific tasks, reminding themselves to stay rational, and working and communicating with other astronauts as much as possible — the opposite of what Kuczaj and I are doing now. As paying passengers, we have no responsibilities aboard *Idabel*, and it's too loud in the front sphere to hold a conversation without yelling. Instead, each of us is stuck in his own thoughts. I wonder what awful thing could happen next and, after thirty minutes of claustrophobic misery, I'm having a hard time staying rational. Kuczaj's jaw is clenched and he looks dazed.

Russian cosmonaut Vasily Tsibliyev had this same problem. In 1997, after four months aboard the space station Mir, he had grown neurotic and depressed. While guiding an unmanned supply ship to Mir's bay, he suddenly got confused and nearly demolished the entire craft. Two years later, two other cosmonauts randomly broke into a bloody fistfight and allegedly tried to sexually assault a female crew member. She went to another room and sealed the door.

It wasn't the workload that got to any of these cosmonauts; it was the claustrophobia of being confined with another human in such a small space. Russian cosmonaut Valery Ryumin once said, "All the necessary conditions to perpetrate a murder are met by locking two men in a cabin of 18 by 20 feet . . . for two months."

I don't have any particularly strong desire to murder, punch, or sexually assault Kuczaj, but then again, we've been confined in this steel pod for only about thirty minutes. We've got three and a half more hours to go before we'll see sunlight . . . or a bathroom. And it's getting colder too. The water temperature is 45 degrees, and the sub's interior hull is chilly to the touch. A sheen of moisture covers the walls and windows.

Stanley announces that we've just passed eleven hundred feet. Suddenly, there's a bright flash on one side of the sub, then another, then two more. Stanley has just snapped on *Idabel*'s eleven headlights. The water in front of us glows as white as milk, then, as our eyes adjust, softens to green-gray, the color of an old television screen. Outside the window, thousands of white flakes stream past us.

Stanley tells us this is detritus from the sunlit waters above. In the ocean, anything that doesn't float must sink: plankton skeletons, fish feces, sloughed-off skin — whatever. It all eventually ends up dissolving into smaller bits and falling to the seafloor in an endless swirl.

The ocean depths suck up not only all the trash but carbon dioxide as well. Phytoplankton, the microscopic algae that make up at least half of all biomass in the ocean, absorb about one-third to one-half of all $CO_2$ and produce more than 50 percent of all the Earth's oxygen. As the oceans warm, phytoplankton will die off. Carbon dioxide levels will rise and oxygen levels will fall.

From 1950 to 2010, the number of phytoplankton species dropped 40 percent — an astounding number. As phytoplankton continue to die off, it will become increasingly difficult for animals on Earth to breathe.

As we cut into deeper water, the detritus flies by faster, like an underwater meteor shower.

"This is incredible," says Kuczaj. He turns on his camera to take some photos. Just as we're scratching the surface of awe, staring at the Milky Way–like light show outside the observation window, Stanley unexpectedly shuts off the lights.

I ask him if he can flick them back on; we're not done getting our minds blown.

"Keep looking out the window," he replies. "Just keep looking."

Without electric lights, it's black out there. I glance at the depth gauge and notice that we've just pushed down past 1,700 feet, the zone where no sunlight can reach. The bathypelagic.

"You see it?" Stanley says. "There, up to the left."

Maybe forty feet away, it looks like fireworks are exploding in the night sky. Then another explosion of light pops below us.

Then more to the right. The colors are brilliant—white with flashes of pink and purple and green. We're looking at what ancient mariners referred to as the burning sea: bioluminescence, the chemical production of light by living organisms. About 80 to 90 percent of ocean life, from bacteria to sharks, use some form of it.

As we look through the front window, the flickers and flashes grow brighter, twinkling and mechanical. A burst of green on the right is matched by a burst of blue a dozen feet to the left. A half a dozen dimmer lights flash in the distance. We can see no shapes, no animals swimming, just flashes of light, like fireflies. We've drifted into a school of . . . something. "It looks like some type of communication," says Kuczaj, raising his camera again.

Bioluminescent animals use light to startle, distract, lure, and communicate. The grotesque-looking anglerfish uses a little light on the top of its head to attract prey. Giant squids—which can grow more than sixty feet long and are believed to inhabit depths even farther down than the bathypelagic—use bright flashes to communicate with other squids, perhaps using something similar to Morse code. The huge eyes of squids, anglerfish, and other deep-sea animals evolved not to process sunlight—they will never see the sun—but to pick up the faintest bioluminescent flickers.

Very little is known about how this light can be used for communication, because so little research has been done with deep-sea animals. Only two giant squids have ever been filmed, and only once have researchers captured a giant squid sending bioluminescent signals.

Still, some terrestrial researchers are drawing on bioluminescence and applying it to their fields. Oncologists are now using the bioluminescent genes of the sea pansy, a gelatinous, jellyfish-like animal, to study how cancer cells and pathogens react to treatments. Sea-pansy genes are also used to research everything from gene expression in stem cells to how viruses infect living organisms.

In January 2000, American artist Eduardo Kac hired a French

genetics firm to splice a jellyfish gene for a green fluorescent protein into an albino rabbit's genome to create the first—and very controversial—glowing-mammal art project. In 2013, an American team with plans to genetically alter plants so that they glow in the dark successfully raised $480,000 on a Kickstarter creative-project funding campaign. The team hopes the glowing plants may one day replace streetlights.

Around the same time that Kac debuted his fluorescent rabbit, scientists added a fluorescent gene to ordinary zebra fish to create GloFish, the world's first genetically modified fluorescent fish. GloFish are now available at pet stores throughout the United States.

STANLEY FLIPS THE LIGHTS BACK ON, and the black water in front of us instantly turns gray with the perpetual snowfall of detritus. The fireworks disappear. The scene becomes somehow stranger. Passing before us is a school of fish, but they aren't swimming horizontally, like normal fish. They're swimming vertically, toward the surface. In the reflection of the headlights, they look like a dozen silver exclamation marks.

While most creatures on land are restricted to a single horizontal plane, those between the surface and the bottom, an area called the midwater, can travel in any direction. This world is utterly featureless and unnervingly constant. There are no mountains here, no skies, no landmarks, nothing to distinguish right from left, up from down. Night never becomes day; there are no seasons. The temperature is almost always the same. There is no specific home for the animals that live here, no place to return to, no destination to reach, just constant drifting. I feel a profound, existential sadness here; it is the blackest, loneliest place I've ever seen.

Danger in the midwater can come from any direction. Some animals escape the monotony by migrating up hundreds of feet into the shallower, lighter waters when the sun is out, then sink back down to camouflage themselves in the black waters at night. This commute is the largest animal migration of life on Earth, and it happens every day. Most animals in the bathypelagic, however, never leave.

IDABEL PASSES TWO THOUSAND FEET. The hull's fizzes and squeaks grow louder and more frequent. The pressure outside is now more than 900 psi. If a pinhole leak suddenly appeared in a wall, the stream would cut through flesh like a scalpel until the stream got bigger and the *Idabel's* walls collapsed. Death at this depth wouldn't happen slowly; we'd be crushed in an instant.

Oddly enough, I feel comfort in this. Before the dive, I expected to feel panicked and stressed at these depths, but now, beneath two thousand feet of seawater, I feel calm, almost serene. Absolutely nothing is in my control—I can't get off, I can't stop the walls from caving in. There's no use complaining or worrying about what will happen next.

It reminds me of a passage from George Orwell's *Down and Out in Paris and London,* in which Orwell, having just been fired from a job washing dishes at a restaurant in Paris and entirely penniless, describes the joy of suddenly reaching rock bottom. "It is a feeling of relief, almost of pleasure, at knowing yourself at last genuinely down and out. You have talked so often of going to the dogs—and well, here are the dogs, and you have reached them, and you can stand it. It takes off a lot of anxiety."

Resting my chin on open palms, listening to *Idabel's* steel frame groan and creak, I realize that if we all die down here, nobody will know what happened. Not even us.

I credit part of my relaxation to Stanley. On land, he was quiet and cagey. He ignored my questions and seemed bothered by my presence. This didn't come as a surprise; he's infamous around Roatan for his prickly temperament.

But down here, thousands of feet below the surface, he's a changed man. He's talking, laughing, and tapping his feet to disco and jazz tunes blasting out of a car stereo he rigged up behind us. We're cruising around in a submarine he built with his own hands, and on his own dime, in a realm where he's spent more hours than anyone else. We're guests in his house, and he seems determined to show us a good time.

We sink deeper. A pale light appears before us. I look through the convex window, and it seems as if we're approaching some distant moon. Details of an alien world gradually come into view. Stanley edges closer, slows down, then turns the *Idabel* so that our window is parallel to the seafloor. We ready ourselves for a landing. Kuczaj and I take a deep breath. *Idabel* squeaks and burps. We've just touched down 2,200 feet below the surface.

The steel walls are now freezing, and the inside temperature has fallen to 65 degrees. Kuczaj and I reach down and pull on our shoes to keep our feet from going numb against the floor. The view outside is lunar — boulders, shallow craters, and broad, open plains, all glowing as white as if the place has just been dusted with snow. But it's not snow; that powdery blanket is the leftover calcium and silicon from billions of microscopic skeletons, a fine substance that biologists call ooze. Because there is no sun to melt it, no wind to blow it, no rain to wash it away, the ooze just stays here, building up about an inch every two thousand years.

We've just landed on the Earth's oldest graveyard.

It seems impossible that anything could survive down here. And yet, all around us there is life, of a variety that's more strange and ugly than I could have imagined.

Moving across the white expanse is a reddish, eel-like fish about two feet long. It waddles by on two stumpy legs. This fish, which neither Kuczaj nor Stanley recognize, looks like an offshoot of the evolutionary tree. To our eyes, it seems to be picking an erratic, drunken path along the ocean floor, and we can't help but chuckle as it passes.

Farther away, a fish the size of a lap dog squats near a rock. Its skin is covered in brown blotches that look like tree bark. Every few seconds it opens its mouth like an old man yawning on a park bench. To our right, a gray shark with a long, ragged dorsal fin drifts sloppily by. It swims in lazy half circles, then stares blankly at us through the window with crossed eyes.

Everything down here seems half developed, awkward, slow-moving, and crippled in some way — failed experiments from God's test kitchen. But this assumption couldn't be more wrong. In a world without light, looks don't matter. What does matter is efficiency and adaptability, and each of these animals, as awkward and ghastly as it might appear, has evolved to fit into its own tiny niche in a harsh environment that would destroy most other creatures.

Food at this depth is extremely scarce. With no sunlight, there is no photosynthesis, and without photosynthesis, no plants, plankton, or any other kind of vegetation can grow. This is a carnivorous world, where animals can survive only by hunting and eating other animals. Muscles and flesh require more fuel and energy than many deep-sea creatures can find. As a result, most have developed gelatinous skin and skeletal frames that are the most efficient design for this deep-water environment.

Movement takes energy too, and so most bathypelagic animals rarely bother. They get their food by sitting in one spot and waiting for unsuspecting prey to come close enough to be eaten. They breed in the same way, simply waiting to bump into a prospective mate. Some animals have increased their chances by becoming hermaphrodites, which allows them to breed with whatever sex might come along. Other animals survive by developing a single heightened sense.

Perhaps the most impressive deep-water adapter is the electric ray, a resident of these parts. This creature should be easy prey: it has poor eyesight and worse hearing. Some electric rays can hardly swim, and others have no teeth. And yet electric rays are some of the most feared predators of the ocean.

In the past few hundred million years, these odd, disk-shaped fish (there are roughly sixty species) have evolved organs that can emit a shock of more than 220 volts — about twice that of a light socket in an American house. This isn't some supernatural power;

all organisms function through a series of electrical discharges, what is known as bioelectricity. The electric ray simply developed organs that maximized its lethal potential.

Humans share this electricity. Every cell in your body contains an electrical charge. Any time you look at something, hear a sound, feel, taste, or think, a storm of electrical discharges explodes inside your cells, going back and forth from your brain to different areas in your body at four hundred feet per second.

This electricity travels by way of a series of circuits called ion channels, tiny proteins in the membranes of cells. These channels can permit or block the flow of electrically charged ions through them.

Think of your nerves as rivers, and your brain as a lake into which all those rivers empty. Ion channels work like little dams to control the flow and direction of signals to and from the brain. You have somewhere in the neighborhood of thirty-five trillion cells in your body, each with its own ion channel, opening and closing in synchronicity to give you a sense of the world around you. A few billion just went off while you were reading this sentence.*

When a nerve fires an impulse, a significant amount of electricity is produced. According to Oxford University geneticist and author Frances Ashcroft, the electric field through the ion channel is the equivalent of about 100,000 volts per centimeter.

A human body generates around 100 millivolts (a measure of potential energy). If all the electricity in a person's body could be harnessed and converted to light, the human body would be sixty thousand times brighter than a comparable mass of the sun.

---

* Computers, which were created before scientists understood ion channels, work on the same principle. The binary code on which all computers are based – 0s and 1s – run in series along binary strings, much the way ion channels run along nerve endings. It's those strings of 0s and 1s that make up every color, sound, movie, song, and program – everything you see and hear – on your computer, the same way the binary of open and closed ion channels are used for all the processes in your body.

Pound for pound, you could be brighter than the brightest star in the solar system.*

Some pharmaceuticals work by closing or opening ion channels, which can allow certain cell functions to return to normal. Ashcroft has written extensively on ion channels and helped pioneer the use of medications like sulfonylurea for ailments like neonatal diabetes. Sulfonylurea treats this disease by closing the defective ion channels in cells that, when open, inhibit the production of insulin.

In Chinese medicine, the body's electric energy is referred to as *chi;* the Japanese call it *ki,* and Indians know it as *prana.* The medical traditions of these eastern cultures are, in large part, based on adjusting the amount of energy in certain areas of the body to promote or restore health.

One of the most striking examples, and perhaps the closest human analog Western science has seen to electric rays, are the Tibetan Buddhist monks who practice the Bön tradition of Tum-mo meditation. These monks can raise the temperature in their extremities by as much as 17 degrees, and they can dry wet sheets on their backs in ambient temperatures of 40 degrees. Their power is nowhere near as great as the electric rays', but it is a clear indication of a capacity we share with these creatures.

STANLEY REVS IDABEL'S MOTOR and we drift a few feet off the sloping seafloor to even greater depths. If we were to keep heading down, we'd eventually reach 25,000 feet. The depth gauge now reads −2,550 feet.

In the distance, a group of glittering disco balls hangs a few feet above the seafloor. It's a school of squids, Stanley tells us.

* At least, according to *Discover* magazine blogger Phil Plait's calculations. Follow his figures (if you can): The sun's volume is $1.4 \times 10^{33}$ cubic centimeters. Each second, each cubic centimeter of the sun emits 2.8 ergs (an erg is a unit of energy). So, the total luminosity of a cubic centimeter of sun is 2.8 ergs per second. The human body has a volume of about 75,000 cubic centimeters. Dividing human luminosity ($1.3 \times 10^{10}$ ergs/sec) by volume gives you 170,000 ergs per second per cubic centimeter.

Each is wrapped in a Technicolor coat more sparkly and gar-
ish than the next. Beside the squids are other animals — jellyfish,
I think — that emit bright pink and purple light. It's like we've
stumbled into some underwater Studio 54.

"Hey, take a look at this," Stanley says as he cranks *Idabel* to
the left. Kuczaj and I crane our necks to get a bit closer to the
front window. The steel walls of the observation deck are freezing now, and droplets of frigid water fall onto our heads and down
our necks.

Stanley stops the sub. A two-foot glob of flashing color approaches, then hovers a few inches from the window. Along the
top of this glob is a blanket of lights, all blinking, one after the
other, in perfect synchronicity. First, only blue lights flash, then
only red; then purple; then yellow, until every color in the spectrum has appeared. Then all the colors flash at the same time
and the spectacle repeats. The hundreds of rows of little lights
are evenly spaced around the glob. It looks like a cityscape at
night: when the lights are red, they look like the taillights of cars
on a freeway; when they're white, they look like a grid of street-
lights as viewed from an airplane thousands of feet above. Between these lights, there is nothing — no visible flesh, no nerves,
no bones or body.

"What the hell is . . ." Kuczaj says, his eyes and mouth open
wide.

Stanley says it's a comb jellyfish, the biggest he's ever seen.
Comb jellyfish, members of the Ctenophora phylum, are common
in deep waters. They propel themselves with an outer layer of
fine hairs called cilia and can grow up to five feet in length. Like
all jellyfish, the comb jellyfish has no eyes, no ears, no digestive
system, no muscles. The orb we're looking at is composed of 98
percent water and a scant network of invisible nerves and colla-
gen, all held together by two layers of transparent cells. It has no
brain, and yet this animal hunts prey, mates, and can move nimbly
through the water.

And there it is, this thing, two feet from our faces, at a depth

DEEP

equivalent to twice the height of the Chrysler Building, watching us with its non-eyes, communicating with its non-brain, and dazzling us with its Las Vegas lights.

THE CTENOPHORA, THE WALKING FISH, the shoal of glittering squids, the vertical feeders—all seem to me like outlandish rarities, but in fact they represent the norm here. The bathypelagic and the sunless depths below house 85 percent of the ocean's life, the largest living space on the planet. There are an estimated 30 million undiscovered species in the ocean but only 1.4 million known species on land. The largest animal communities on the planet and greatest number of individuals live below three thousand feet.

As I'm sitting in this cramped metal sphere peering through the window at a seldom-seen habitat, I feel an emptiness in my chest that breath can't fill. This is the real Earth, the 71 percent silent majority. And this is how it looks—gelatinous, cross-eyed, clumsy, glowing, flickering, cloaked in perpetual darkness and compressed by more than a thousand pounds per square inch.

The azure sphere we see from the space is only a veneer. Our planet isn't really blue, it's not filled with leaves of grass, clouds, color, and light.

It's black.

# −10,000

THE RIDE IN STANLEY'S YELLOW submarine, however wondrous, only delays the inevitable: the misery of training to freedive. I have eight weeks before Schnöller's sperm whale mission in Sri Lanka. I can't go if I can't freedive with the team. And so I practice. A lot.

Training for deep dives isn't an option in San Francisco's open waters — the visibility is poor, the water frigid, the tides deadly, and there's the ever-present menace of great white sharks. Instead, I focus my efforts on pool and surface training. A few times a week, I throw my wetsuit and mask into a backpack and bike down to a local public pool to swim underwater laps beneath the dangling feet of elderly women. The lifeguard, who I later learn is a freediver himself, keeps a watchful eye on me. After a few weeks, he takes it upon himself to start coaching me, Mr. Miyagi–style.

His torture device of choice is an orange safety cone that he moves along the edge of the pool, forcing me to hold my breath a few seconds longer with each progressive dive. Improvement in this drill is measured by horizontal distance instead of time spent underwater. I call it Subaquatic Schadenfreude, because making these longer dives isn't easy, and the lifeguard knows it. He chuckles when I resurface, flushed in the face, gasping for air, and looking around bleary-eyed while flapping my numb hands in an effort to restore circulation. Aches, pains, numbness — these are asphyxiation's calling cards. He'd experienced them too. Every freediver in training has.

The drill works. After a month, I double my underwater distance, from about seventy-five to a hundred and fifty feet.

During days off from pool training, I practice static breath-holds while splayed on a yoga mat in my living room. Dry runs are no more tolerable than wet, but they serve a unique purpose: they help me get used to carbon dioxide buildup in my body.

That nagging, need-to-breathe feeling you get holding your breath is triggered not by oxygen deprivation but by buildup of $CO_2$. Comfort with this buildup is what separates good freedivers from great ones, or good ones from guys like me. Freedivers condition their bodies to tolerate high levels of $CO_2$ using timed breath-hold exercises called static tables. Essentially, it's interval training. Breathe two minutes, take four huge breaths, hold breath for two minutes; breathe one and a half minutes, take four huge breaths, hold for two and a half minutes, and so on.

The aim of static tables is to increase breath-holding time while decreasing the rest interval. Within a few weeks, I hit my goal of three-minute breath-holds with only one-minute rests in between.

THERE WAS ANOTHER, SELDOM-DISCUSSED side effect of static training that went beyond increasing $CO_2$ tolerance: it gives you a bone-deep high. This high falls somewhere between the endorphin rush of intense exercise and the dirty, intoxicated feeling you get from drinking bad alcohol in a hurry. A warm spaciness takes over and you feel the electric pulses of your nerve endings firing through your entire body, or you're at least high enough to imagine that something like that is happening. Your mind wanders to happy places.

I begin practicing static breath-holds in different locations around the house. Schnöller warned me that if I ever did this (and almost every freediver in training *does*), I should be sitting or lying down and have nothing sharp nearby. Blackouts can happen on land as easily as in the water, and sometimes it's hard to know exactly when they're going to hit. One moment you're holding your breath, doing the dishes, and feeling great. The next you're unconscious on the kitchen floor in a pool of your own blood. That's exactly what happened to one of Schnöller's friends. You'll stay unconscious anywhere from a few seconds to around a minute. Your brain will eventually wake itself up, discover that the rest of the body is not really underwater, and then trigger your lungs to inhale. Blackouts on land are harmless as long as you've landed in a soft spot.

I have one close call. A few weeks into my static training, I try to spice up some boring office work by attempting consecutive three-minute breath-holds. I don't realize anything has happened until I find my head hanging low, one arm dangling at my side, and hot tea spilled all over my keyboard. I'd conked out — just for a second, it seems. I never felt like I was about to be unconscious; it was a seamless transition from pre-unconsciousness to post-unconsciousness. I deduce that something had happened only from the changes in my environment. It creeps me out.

Despite that near-miss, I don't confine breath-hold training to the safety of my home.

One of the best surface-training methods is so-called walking apnea, which involves holding your breath and walking over a soft surface (in case you pass out) for extended distances. The idea is that the oxygen your muscles use when you're walking slowly is about the same amount of oxygen muscles use during a freedive. You start by holding your breath while standing still for about thirty seconds until you feel your heart rate decrease, then you walk slowly in a straight line, turn around when you feel you've reached your halfway point, and walk back to your starting place. The distance you travel is about how far you'd be able to hold your breath during a deep dive.

After a month of constant practice, I can easily walk more than two hundred feet (a hundred feet each way) without breathing.

But freediving is more than just walking and holding your breath. My greatest challenge, as is true for many beginners, is learning how to equalize my sinuses (or pop my ears) thoroughly and in quick succession. No matter how hard I tried to do this during my dive attempts at 40 Fathom Grotto, I couldn't seem to perform them fast enough to attain any real depth. The simple explanation: I was doing it all wrong.

When the average person tries to equalize his ears, he puffs out his cheeks and blows hard, so that compressed air enters the sinus cavities that lead to the ears. This method, called the Valsalva maneuver, is used by about 99 percent of the population, and it's usually effective. But it doesn't work when you're freediving past around forty feet. As you dive deeper, air becomes more and more compressed in the lungs, until there isn't enough left to push into the ears. The Valsalva method becomes useless.

Most freedivers and some jet pilots (who need to equalize quickly during ascents and descents) use the Frenzel method, which traps air inside the closed circuit of the sinus cavities and allows for immediate and thorough releases of pressure. This method is complicated and many people do it wrong, which can cause serious problems at depth. I hire Ted Harty, the team captain for the U.S. freediving team, to lead me through a thirty-minute training session on Skype. (When the lesson starts, I quickly recognize Harty as the guy with the gill tattoos on his ribs who, months back, had monitored me during my four-minute breathhold at the Performance Freediving International course in Tampa.)

"The big difference between Valsalva and Frenzel," Harty begins, "is that in Valsalva, the throat stays open; in Frenzel, it's shut."

Over ten minutes, he guides me through some exercises that include coughing a *T* sound and groaning with my mouth shut. Both act on the epiglottis, the fleshy flap that covers the windpipe, so that I can open and shut it at will. Next, Harty shows me how to "puke" air up from my stomach and "jackhammer" it with my tongue into the sinus cavities. By trapping air in my head (instead of the Valsalva method of pushing it up from the lungs), I'm able to shuffle air back and forth between the sinus cavities and release pressure in a fraction of a second. Once I get it, it works every time.

The maneuver is as awkward as it sounds and nearly impossible to explain if you don't have someone showing it to you, which is why Harty offers his private Skype sessions. It also takes a lot of practice. Harty tells me to repeat the Frenzel method at least three hundred times a day for the next week and then use it during my next pool-training sessions. Before he signs off, he offered a final piece of sage advice.

"Remember: never, ever freedive alone," he says. "I have students who sign up for courses. But they never show up. You know why?" He pauses. "Because they died practicing alone. Don't ever do it."

We hang up, and I head out to the park to walk my dog, holding my breath and puking air into my head the whole way.

---

The one part of me that still needs to be trained for the rigors of freediving is my mind. For assistance in this realm, I turn to Hanli Prinsloo, a former competitor turned spiritually minded freediver. As with many former competitors, she gained wisdom only after skirting death.

"I felt this irritation in my throat," says Prinsloo. "I coughed and there were flecks of blood."

I'm sitting at a worn wooden table with her, inside a crowded restaurant in Kalk Bay, a trendy former fishing village about twenty miles west of central Cape Town, South Africa. Prinsloo, who lives just up the street, is wearing a thin black down jacket, jeans, and lamb's-wool-filled boots. Behind her is a large window, and looking through it, I see the slick backs of southern right whales bend and push beneath an envelope of gray ocean. Anywhere else, this would be a million-dollar view, but here whales are about as common as dogs on the beach, at least in springtime. Prinsloo is framed in the center of my view, sipping wine and laughing as she describes how her larynx ripped apart.

"I wanted to see how far my body could go," she says. "You know, test my limit."

The dive Prinsloo is describing happened in August 2011, a month before I met her at the World Freediving Championship in Greece. She was training in Dahab, Egypt, with her friend Sara Campbell before an attempt at a women's world record in the discipline of constant weight (CWT). The women's CWT record was 203 feet at the time; Prinsloo planned to increase it to 213.

For months, she followed a rigorous training schedule: diving with half-filled lungs to around 120 feet several times a day, doing yoga, practicing static breath-holds. She ate a raw vegan diet — no wheat, sugar, or alcohol — in order to boost the oxygen stores in her blood and cut down on excess mucus, which would make it difficult to equalize quickly at depth.

On her first practice dive of the day, she took one last gulp of air, turned, closed her eyes, and kicked her way down the rope.

"Right at the beginning, I felt that this dive was different," she says. "I was exhausted and tense, not feeling myself." Prinsloo ignored the warning signs and forced herself to go deeper. At around 130 feet, she felt a contraction in her stomach. She rarely felt that, and never on the way down. She still had more than 200 feet to go.

Prinsloo managed to complete the dive and return to the surface still conscious. She exhaled the stale air from her lungs, took a big inhale, then coughed. Droplets of blood shot from her mouth. Her larynx had been torn under the pressure.

Ordinarily, the larynx can withstand the stresses at extreme depths, but only if the body is relaxed. If a diver tenses up, the soft tissues can rupture, sometimes causing serious or permanent damage, sometimes resulting in death. Sara Campbell shared a dire prediction with her. "She said I was lying to myself. I was so untrue that I was starting to hurt my own body." For Prinsloo, it was a turning point.

Prinsloo dropped her world-record bid and officially ended her thirteen-year competitive-freediving career. Two months before coming to Dahab, she had traveled to Dharamsala, India, and lived for five weeks in a Buddhist temple, where she meditated for twelve hours a day, practiced yoga, read philosophical books, and, in her words, "spent one month just breathing." At the end of her stay, she rediscovered a "stillness" in herself. It was the same stillness that had first attracted her to freediving fifteen years earlier, but it had been lost in her ambition to keep going deeper.

"In Dharamsala, I remembered that freediving was all about letting go," she says. "After Dahab, I was reminded, again, that you can never force yourself into the ocean. You do that and"— she pauses — "you'll just get lost."

I'D COME TO KALK BAY for six days, hoping that Prinsloo's holistic approach could help me get through the doorway to the deep. I needed something.

My course with Eric Pinon months back gave me all the tools to freedive, but I still didn't know how to use them. My head and ears still ached on surface dives, and without Pinon beside me, my thoughts gave way to paralyzing fear whenever I felt the pressure of deep water below twenty feet. Then I'd immediately imagine the blacked-out faces of divers I saw in Greece. This sounds melodramatic, I know, but it's true. Those dead eyes and bloated necks were powerful images — some of the most gruesome I'd ever seen. They returned to me inevitably every time I dived. My thoughts would snowball. I'd then imagine myself blacking out, turning blue. I'd lose concentration, feel an incredible urge to breathe, and scramble back to the surface for air. My dive watch would show a mere twenty seconds, most of it spent in anxious misery.

Prinsloo is a world expert in the art of letting go. The work keeps her busy. Last month, she tells me, she was hired by the Springbok Sevens, a rugby team in Cape Town. "Some of these guys were scared of the water; they didn't even know how to swim!" she says. A few weeks later, they were swimming underwater laps.

"CAN YOU HOLD THIS FOR a sec?" Prinsloo says, handing me a stainless-steel water bottle.

It's early morning the following day. I'm in the passenger seat of Prinsloo's truck, a beat-up baby blue Toyota Hilux that she affectionately calls Freya. Prinsloo is flooring Freya through a Tolkien landscape of five-hundred-foot vertical cliffs and overgrown bushes along the shores of a turquoise ocean. She's speaking Afrikaans into a cell phone she's holding in one hand, steering the truck around hairpin turns with the other, and, between breaths, talking to me in English. "God, it's just been so long since I've been in the water, it drives me crazy," Prinsloo says. She steers with her knees for a moment as I pass her a water bottle. "Six days."

She raps out a few Afrikaans words into the phone, laughs, then turns back to me. "But, for me, that's forever."

In the back seat of the truck is Jean-Marie Ghislain, a fifty-seven-year-old former real estate executive from Belgium who quit his job six years ago after having a life-changing encounter while swimming with sharks. Ghislain now runs a nonprofit conservation program called Shark Revolution and spends nine months of the year trotting around the globe photographing oceanic animals and freedivers, sometimes both of them at the same time.

As we ride, Prinsloo shares some maxims, a kind of Ten Commandments of freediving:

Freediving is more than just holding your breath, it's a perception shift.
Don't kick down the doorway to the deep; slide in on your tiptoes.

Never, ever dive alone.
Always enter the ocean in peace with yourself and your
   surroundings.

At the heart of the list is peaceful coexistence with the water
and its inhabitants, whether they're other freedivers, seals, dolphins, whales, even sharks. Prinsloo demonstrated this yesterday
at the Two Oceans Aquarium in Cape Town when, for a photo op,
she dove headfirst into a five-hundred-thousand-gallon aquarium
filled with ragged-tooth sharks. The sharks didn't attack. Most
looked as though they couldn't care less. The few sharks that took
an interest in Prinsloo let her swim side by side with them, almost
as if they were welcoming her into the shiver. It was fascinating
to watch, but it made my skin crawl.
   Prinsloo thinks my fear of sharks adds to my already lengthy
list of freediving inhibitions. Perhaps she's right. In thirty years
of swimming in the Pacific Ocean, I've had some bad experiences.
I've seen the teeth marks of a great white on the decapitated body
of a seal along the shore of my favorite surf break. I've traced
my finger along surfboards mangled by two-foot-wide bites. I've
seen Frankenstein scars along the stomach of a surfer attacked
by a shark just days after I'd surfed at the same site. I've been to
Réunion Island twice. Yes, I know sharks are an essential part of
the oceanic ecosystem, and I certainly don't want them killed. But
I also have no interest in encountering them in the wild.
   Prinsloo believes that if I confront this fear, if I dive with
sharks and see them for myself, I'll experience the perception
shift she keeps talking about. And that shift may carry over to my
"unreasonable fear" about being unable to breathe underwater, of
not seeing the surface. It's all part of the art of letting go.
   Thirty minutes after we headed out, Prinsloo pulls over at
Miller's Point, a popular freediving spot that's also home to dozens of sevengill cow sharks. Cow sharks, which get their name
from their large and innocent-looking bovine eyes, are considered
a mellow species and not prone to attacking humans (at least, not
often).

We suit up, walk down to the water, and swim toward the horizon until Freya is a speck of blue against the rocky landscape. Below us, morning sunlight cuts through columns of kelp to create what looks like crisscrossing spotlights in the brilliant green water. The visibility is good for these waters, maybe eighty feet.

"You see it?" says Prinsloo, lifting her head out of the water. Below us, a cow shark about the size of an adult human swims by, some twenty feet down on the ocean floor. Prinsloo takes a gulp of air, dives, approaches, and swims beside it. Once she's at eye level with the shark, she kicks in rhythm with the shark's back fin. The shark makes a sharp right turn; Prinsloo follows it, a bit closer this time. Then the shark turns right again, making a large circle. It quickly wiggles its backside back and forth. Prinsloo and the shark are playing with each other.

The shark eventually swims off, but another one arrives a few minutes later, and the scene repeats. This goes on for an hour.

Finally, curiosity trumps fear and I take a gulp of air and dive down about ten feet to join them. The sharks keep their distance; I make them nervous with my awkward movements and constant, hasty ascents to grab air. But they don't swim off. After a while, the sharks and I edge closer.

I'll admit, being with these animals doesn't stir up any great affection for them, but I do feel a thin slice of camaraderie. We're sharing turf. They can devour me, but they don't. I could watch them from a boat, but I'm down here. Perhaps this is all part of them casing the joint before raiding the store. Or perhaps that's just my "unreasonable fear" talking again.

After a while, I just stop thinking about it and swim with the sharks.

---

The next few weeks, I work with Prinsloo to develop a regimen. I read the *Manual of Freediving,* a 362-page bible of the sport, cover to cover. I scour the Internet, watching countless YouTube instructional videos and reading freediving blogs. I practice and practice. I tell myself and Schnöller that I'm ready.

A MONTH LATER I'M IN the passenger seat of a white van, on a dusty, potholed road somewhere along the northeastern coast of Sri Lanka. It's 9:00 p.m. And the stars are out. "Is this the right way?" I ask our driver.

He's a local named Bobby; that isn't his real name, but that's what he wants me to call him. Bobby is shaking his head and flashing me a reassuring smile. It's the same smile he used ten minutes ago when he took a wrong turn into someone's front yard, the same one he gave me twenty minutes before that when he brought the van to a dead stop in the middle of a two-lane freeway, stepped out into oncoming traffic, and ran across the street to ask a barefoot man on a bike for directions.

"Bobby? Is this the *right* way?" I repeat.

That smile.

Then Bobby suddenly pulls into a driveway. Through the headlights, it looks like we've just entered a junkyard. Guy Gazzo, the seventy-four-year-old freediver from Réunion, mumbles something in French in the seat behind me. Gazzo is sitting beside Diderot Mauuary, an acoustic scientist from northern France. Trailing us in an identical white van is Fabrice Schnöller and an American film crew.

We've all just spent twelve hours driving through steep mountain roads, jungles thick with elephants, and dusty towns filled with men in baggy slacks selling boiled peanuts and green bananas. Now we're two hours late and we're all getting irritated.

"Bobby?"

He pulls out of the driveway and takes a left. This road is narrower and bumpier. Bushes scrape the doors. The eyes of unknown animals glow from copses of coconut palms. A dog barks. Bats the size of rats flutter and swoop inches from the windshield.

Minutes later, we come to a stop in a barren sandlot. To the right is a creepy-looking, three-story pink-concrete building. A single, bare light bulb shines over a white plastic table on the patio, giving the scene an Edward Hopper feel. Bobby exhales, pulls the key from the ignition, and smiles. We've arrived at our destination, he says: the Pigeon Island View Guesthouse.

I'm in room 6, up three flights of stairs, near the back of the building. In one corner of my room, beside a lime-green wall, is a bed so short that my legs will dangle off the end at midcalf. Overhead, a ceiling fan swings precariously from two wires, its blades whipping clumsily around like a helicopter about to crash. A pink net covering the bed is supposed to keep flies and mosquitoes out but has done little to stop the fleas, which hop like popcorn on the sheet and pillowcase. I'll be sleeping here for the next ten days.

I'VE JOINED THE DAREWIN TEAM in Trincomalee, a Podunk town along the northeast coast of Sri Lanka, to swim with sperm whales, the world's deepest-diving animals.

Sperm whales can go down as far as ten thousand feet, but studying them at such depths is impossible: few submarines or ROVs could make it, and they wouldn't see anything if they did. No sunlight gets down there, and the artificial lights of a sub would scare off the whales.

As with sharks and dolphins, the best way—the *only* way—to film and study sperm whales is at the surface.

Hunting down whales and forcing your presence on them never works—they get spooked and dive; they swim away; or they attack. The whales must choose to come to you, and they'll choose a freediver more often than a boat, a scuba diver, or a robot.

What attracts the whales to Sri Lanka is Trincomalee Canyon, an eight-thousand-foot-deep chasm that stretches twenty-five miles across the Indian Ocean, from the northern tip of the country into the Trincomalee harbor. Sperm whales come here to feed on deep-water squids, socialize, and mate during their annual migrations from March through August. They've done it for as long as anyone can remember, and probably for millions of years before that.

Unlike other deep-water chasms and sperm whale hotspots, Trincomalee Canyon is close to shore, so it's easy to run trips out to the whales by day and return to land at night. This will save our team from hiring a live-aboard research vessel, which can cost thousands of dollars a day. But the big attraction is that there are no permits to obtain, no authorities to evade, nothing to stop us from freediving with the whales. Because there's nobody here.

During the Sri Lankan civil war, which lasted from 1983 to 2009, separatists led by the Liberation Tigers of Tamil fought the Sri Lankan army for control of the northeast coast. Trincomalee was a war zone. Few tourists came here; what little infrastructure existed was quickly destroyed, and the area took a direct hit during the Indian Ocean tsunami of 2004. The unintended benefit for sea life was a coastline that went years without any significant human presence. Cruise ships have never motored through here, and there is no whale-watching industry to speak of. In many ways, the waters of Trincomalee look about the same as they did thousands of years ago.

Today, it's one of the world's best places to see and study sperm whales.

COMING HERE was my idea. After visiting Prinsloo in Cape Town, I connected her with Schnöller and the two proposed that we all go out to Trincomalee on a freediving sperm whale–research expedition. A few months later, tickets were purchased and travel arrangements made. Somehow, at nine thirty, after days of air travel from five different points of the globe, we are all sitting together on the patio of the Pigeon Island View Guesthouse. On one side of a patio table is the DareWin crew: Schnöller, Gazzo, and Mauuary. On the other is Prinsloo's team. She's brought her new boyfriend, a six-foot-two aquatic he-man from Los Angeles named Peter Marshall. Marshall broke two world records in swimming at the 2008 Olympic trials. Beside him is Ghislain. He tells me that after we met in Cape Town, he went to Botswana to swim with crocodiles. The trip ended after the first day when a team member had an arm eaten off.

Also in the group are three members of an American film crew who have come to shoot footage for a planned documentary on Schnöller's work with dolphin and whale clicking communication.

Thirty years ago — to the week — another American film crew came to Trincomalee and captured the first footage of sperm whales in their natural habitat. The resulting film, *Whales Weep Not,* narrated by Jason Robards, became an international sensation and helped spark the Save the Whales movement.

Our crew hopes to have a similar impact by capturing the first 3-D footage of sperm whales and human-and-whale freediving interactions. DareWin scientists will use the click data collected from various hydrophones on the cameras to help decipher what they believe is a sperm whale click language.

But for any of this to work, we'll need to find some whales.

DURING MY DIVE IN STANLEY'S sub to 2,500 feet, I felt more remote from the world I knew than I ever had before. And the gelatinous, awkward, eyeless, and brainless creatures at the bottom of the Cayman Trench seemed farther from humanity than anything I could have imagined.

I presumed that this sense of remoteness would only intensify as I investigated deeper realms. Sperm whales would seem to buttress that view.

These creatures look nothing like us. They weigh up to 125,000 pounds and lack the limbs and hair of land-dwelling mammals. Their insides are as unlike ours as their outsides are. The sperm whale has four stomachs, a single nostril on top of its head, and a three-hundred-gallon reservoir of oil that gives its enormous nose its distinctive shape. They can hold their breath for up to ninety minutes at a time and dive to depths of 10,000 feet. Yet, in two related and crucial ways — language and culture — sperm whales more closely approximate human culture and intellect than any other creature on the planet.

"It's sort of strange. Really, the closest analogy we have for it would be ourselves," said Hal Whitehead, a Canadian biologist who has researched sperm whales for thirty years. Whitehead was referring to elaborately developed sperm whale groupings, which he called "multicultural societies." Within these societies, whales communicate in dialects and share behaviors distinct from other whales who live nearby.

Each sperm whale society is made up of tight-knit family units of "nursery schools" that contain from ten to thirty mature females and their male and female offspring. Calves are raised not just by their mothers but by an entire matriarchal group of relatives, which includes aunts and grandmothers. Females stay in these families their whole lives, while males, called bulls, are taught at an early age to become more independent. By their teens, bulls join groups, or gangs, of other bulls and wander the ocean looking for food – and sometimes for trouble. Bulls will eventually strike out on their own to live bachelor lives in the Arctic and Antarctic Oceans, visiting the equator – "for summer vacation," said Whitehead – every spring to mate and socialize for six months before returning to their solitary winter homes.

Sperm whale clicks—which are used for echolocation and communication, and max out at 236 underwater decibels—can be heard several hundred miles away, and possibly around the globe. Sperm whales are the loudest animals on Earth.

In air, a 236-decibel sound would be louder than two thousand pounds of TNT exploding two hundred feet away from you, and much louder than a space shuttle taking off from two hundred fifty feet away. In fact, 236 decibels is so loud that a sound of that intensity cannot exist in air. Above 194 decibels, sound waves turn into pressure waves.

Although decibels are used to measure sound intensity in both air and water, the intensity of sound in both of these mediums differs. In other words, a decibel is a relative unit of measure. (For instance, an 80-decibel sound will be quieter in water than it is in air.) But this distinction doesn't make sperm whale clicks any less impressive, or lethal. Even underwater, sperm whale clicks are so loud they could not only blow out human eardrums from hundreds of feet away, but, some scientists estimate, vibrate a human body to death.

The extraordinary power of clicks lets whales use them to perceive a remarkably detailed view of their environment from great distances. They can detect a ten-inch-long squid at a distance of more than a thousand feet and a human from more than a mile away. Sperm whales' echolocation is the most precise and powerful form of biosonar ever discovered.

The sperm whale's brain, like its clicks, both distinguishes it from a human's and suggests surprising similarities between the two species.

Six times the size of the human brain and in many ways more complex, the sperm whale's brain is the largest brain that's ever existed on Earth, as far as we know. The inferior colliculus in the sperm whale brain, which helps sense pain and changes in temperature and serves as an auditory pathway from one area of the brain to the other, is twelve times larger than a human's; its lateral lemniscus, which processes sound, is two hundred and fifty times the size of a human's. The neocortex, the part of the brain that, in humans, governs higher-level functions such as conscious thought, future planning, and language, is estimated to be about six times larger in the sperm whale brain than it is in ours.

It's also possible that whales have emotional lives not unlike our own. In 2006, researchers at New York City's Mount Sinai School of Medicine discovered that sperm whales had spindle cells, the long and highly developed brain structures that neurologists associate with speech and feelings of compassion, love, suffering, and intuition — those things that make humans *human*.

Sperm whales not only have spindle cells, but have them in far greater concentration than humans do. Scientists believe these cells evolved in sperm whales more than fifteen million years before they did in humans. In the realm of brain evolution, fifteen million years is a very long time.

"It's absolutely clear to me that these are extremely intelligent animals," said Patrick Hof, one of the researchers who made the discovery.

It's this brain, specifically the oversize neocortex and spindle cells, that has brought Schnöller and the DareWin team to Sri Lanka.

A nonscientist might call love, suffering, and compassion the stuff of poetry. And no poetry was ever conveyed without words or something like them.

OUR FIRST TWO OUTINGS are a disaster. Both days, we spent several hours in two tiny, shadeless fishing boats juddering around the ocean without seeing any whales. The film crew's cameraman got seasick the first day and refused to go back to sea. Without a cameraman, and still without any usable footage, the director threatened to pull the plug on the documentary.

On the evening of the second day, I meet Schnöller on the second-story patio. He's sitting alone, haloed in mosquitoes. The blue fluorescent beam of a headlamp shines down on a table filled with half-assembled underwater-camera casings. Behind him, a waxing moon hangs low over a black sea.

"This is very hard work, you see," he says, looking up as I take a seat at the table. He's wearing an American flag headband and knockoff Facebook sandals that he picked up at a junk store on the way here, and he looks as ridiculous as that description makes him sound. "Ocean research takes patience, lots of patience, persistence, and is very physically exhausting."

Schnöller grew up in the west African nation of Gabon, the son of a former French army lieutenant who worked for then dictator Omar Bongo. The family's house was located beneath a canopy of mango trees at the shoreline of an unpopulated beach, which was where Schnöller spent much of his youth. He told me earlier how he remembered watching crocodiles from a nearby river crawl up the front porch and eat food from the dog bowl. Sometimes while the family was eating dinner, giant mambas would slide in through wooden planks in the roof and drop down on the dining-room table. Schnöller's father kept a shotgun close by, and after a few years, the roof was peppered with holes.

On weekends, Schnöller would sail along Gabon's wild coast

and make camp on unexplored islands. He learned how to navigate through the ocean's many moods, keep cool in crises, and improvise his way out of trouble.

Schnöller knows that Prinsloo's team has been criticizing him and that the film crew is about to leave, but he shrugs it off. "There are no fast results in this research," he says. "That's why so few people bother doing it."

Actually, he corrects himself, *nobody* is doing it.

Of the twenty or so sperm whale scientists in the field, none dive and interact with their subjects. Schnöller finds this inconceivable. "How do you study sperm whale behavior without seeing them behave, without seeing them communicate?" He's convinced that to understand sperm whales, one must first understand their communication, and to understand their communication, one needs to understand their language, which he believes is transmitted through clicks.

"These patterns are very structured; this is not random," he says, taking a sip of beer.

SPERM WHALES PRODUCE FOUR DISTINCT vocalization patterns: normal clicks, for tracking down prey at distances of more than a mile; creaks, which sound, despite their name, like machine-gun fire, for homing in on close-range prey; codas, the patterns used during social interactions; and slow clicks, which no one quite understands. One theory is that bulls use slow clicks to attract females and scare off other males. The clicks are very similar to dolphin clicks but more complex.

Coda clicks, the focus of Schnöller's work, are used only during socializing and are significantly different from clicks used to aid perception and navigation. They sound unremarkable to the human ear—something like the *tack-tack-tack* of marbles dropped on a wood table. But when the clicks are slowed down and viewed as a sound wave on a spectrogram, each reveals an incredibly complex collection of shorter clicks inside it.

Inside those shorter clicks are even shorter clicks, and so on. The more closely Schnöller focused in on a click, the more detailed it became, unfolding on his computer screen like a Russian nesting doll.

An average click lasts anywhere from twenty-four milliseconds (thousandths of a second) to seventy-two milliseconds. Inside these clicks are a series of microclicks, which themselves are separated by microseconds, and so on. All these tiny clicks inside the coda are transmitted at very specific and distinct frequencies. There could be even shorter, organized click patterns within these microclicks, but Schnöller's machines—which record at 96,000 Hz, the highest speed available on most modern audio equipment—aren't fast enough to process them.

Schnöller tells me that sperm whales can replicate these clicks down to the exact millisecond and frequency, over and over again. They can also control the millisecond-long intervals inside the clicks and reorganize them into different structures, in the same way a composer might revise a scale of notes in a piano concerto. But sperm whales can make elaborate revisions to their click patterns, then play them back in the space of a few thousandths of a second.

"When you think about it, human language is very inefficient, it is very prone to errors," Schnöller says. Humans use phonemes—basic units of sound, like *kah, puh, ah, tee*—to create words, sentences, and, ultimately, meaning. (English has about forty-two phonemes, which speakers shuffle around to create tens of thousands of words.) While we can usually convey phonemes clearly enough for others to understand them, we can never fully replicate them the same way each time we speak. The frequency, volume, and clarity of the voice shifts constantly, so that the same word uttered twice in a row by the same person will usually sound discernibly different, and will always show clear differences on a spectrogram. Comprehension in human language is based on proximity: If you enunciate clearly enough, another speaker of the same language will understand you; if you bungle too many vowels and consonants, or even pronunciation (think of French or a tonal Asian language), then communication is lost.

Schnöller's research suggests that sperm whales don't have

this problem. If they're using these clicks as a form of communication, he believes, it would be less like human language and more like fax-machine transmissions, which work by sending out microsecond-length tones across a phone line to a receiving machine, which processes those tones into words and pictures. Perhaps it's no coincidence that a pod of socializing sperm whales sounds a lot like a fax transmission.

Human language is analogue; sperm whale language may be digital.

"WHY DO THEY HAVE SUCH huge brains, why are these patterns so consistent and perfectly organized, if they aren't some kind of communication?" Schnöller asks rhetorically. He mentions that sperm whales have more brain mass and brain cells controlling language than humans do. "I know, I know, this is all just theory, but still, when you think about it, it just doesn't make sense otherwise."

To illustrate his point, Schnöller relates an encounter he had the previous year with a pod of sperm whales. The pod included both adults and their young and were hanging out in the water, clicking and socializing, when Schnöller approached them with a camera attached to a bodysurfing board. A calf swam over and faced Schnöller, then took the camera in its mouth. A group of adults immediately surrounded the calf and showered it with coda clicks. Seconds later, the calf let the camera go, then backed up and retreated behind the adults without ever looking at them. To Schnöller, the young whale looked ashamed. "It got the message not to mess with us." He laughs. "That's when I knew, they had to be talking to it. There's just no other way."

Schnöller says he's also witnessed, on numerous occasions, two sperm whales clicking back and forth to each other as if they were having a conversation. He's seen other whales pass clicks and then suddenly move in the same direction. He's watched a whale bend its head in exaggerated motions to face one whale head-on and pass one pattern of clicks, and then bend in another direction to face another whale and pass a completely different pattern. To Schnöller, it all looked like communication.

But neither Schnöller nor anyone else will be translating the cetacean language anytime soon. It's too complicated, and both resources and personnel are too scarce to study it closely. The DareWin team has come here to collect data in the hope of simply proving that sperm whales use clicks as some form of communication. They'll record as much sperm whale socializing as they can, then correlate coda clicks with specific behaviors.

That's what the crazy-looking pod is doing at Schnöller's feet. The device, called the SeaX Sense 4-D, is a glamorized underwater-camera housing covered with twelve minicameras and four hydrophones, all placed at different angles. With it, Schnöller can record high-definition audio and video in all directions at once.

Schnöller explains that the sperm whale, like the dolphin, probably processes sound through an acoustic sac found in the upper jaw at the tip of its huge nose, and, like a dolphin's, has thousands of receptors to gather sound. Having more receptors (more ears, essentially) allows the whale a significantly broader and more accurate view of its environment. Using its nose and clicks for echolocation, a sperm whale can "see" clearly in all directions at once.

SeaX Sense 4-D, Schnöller says, "replicates what the sperm whale sees and hears" by capturing 360-degree video and surround-sound audio. An additional, smaller 3-D camera with just two hydrophones will replicate the human experience. Data from these two machines will be uploaded into a software program DareWin engineers developed that can pinpoint which whale clicked at which other whale and at what time. If a whale reacts in a specific way to the same click pattern, that will suggest these clicks are coded with some information. Researchers will then work backward and analyze these clicks for patterns and try to piece together the click vocabulary.

It's not the Rosetta stone, Schnöller admits, but it's a start. Nobody has ever recorded sperm whale interactions and behaviors with such sensitive equipment before, because no such equipment had existed. Schnöller built all this stuff from handouts and scraps.

Freediving with these rigs, he has recorded twenty hours of close-up sperm whale interactions — the largest and most detailed such collection in the world.

AT SEVEN IN THE MORNING on the third day, the boat captains arrive and lead us back to our hired "research vessels" — two beat-up fishing boats with wooden planks for seats.

The remaining two members of the film crew and the Dare-Win team will take one boat; Prinsloo's team will take the other. I'll be alternating between the two. The plan is to head out together, several miles off the coast, to a spot in Trincomalee Canyon where the seafloor drops off to a depth of more than six thousand feet. From there, we'll split up and look for whales. Should someone on either boat spot any, he'll use a mobile phone to alert the other boat. We'll then trail the whales, wait for them to slow down or stop, and get in the water with them. With any luck, they'll approach and interact with us.

We pack up, squeeze in, and set off south toward the horizon, our rickety craft riding low in the water. Hours later, we're twenty miles off the coast, floating in a dead-calm sea. No whales. I'm beginning to side with the film crew: this expedition feels hopeless.

"There were just so many out here last year," Prinsloo says apologetically. She's curled up in a sheet wet with seawater and sweat, leaning against Peter Marshall. Both of them are wearing T-shirts around their faces, so only the lenses of their sunglasses peep through. "I don't know," Prinsloo laments. "I don't know what happened."

Ghislain wipes his sweaty palms against his light blue Abercrombie & Fitch T-shirt. He emits an exaggerated sigh, takes a sip of water, and turns to stare into the open ocean. A minute becomes an hour; an hour becomes two. I check my dive watch: the temperature gauge reads 106. Even my fingers are sunburned.

I recall when Schnöller told me months ago that he sees dolphins and whales only 1 percent of the time that he spends looking for them, and that he films them 1 percent of that 1 percent. Now I worry that the percentage is actually far lower.

I've discovered in the past fourteen months that deep-sea research has less to do with actually researching the mysteries of the sea and more to do with watching Tom Cruise movies on airplanes, brushing your teeth in gas-station bathrooms, sleeping in fleabag hotels, diarrhea, picking dead skin from your peeling shoulders, arguing, eating stale croissants for lunch and dinner, explaining to loved ones that you won't be home anytime soon, and sitting in little boats over deep-water trenches in the middle of nowhere writing sentences like this one on a damp notepad.

Another hour passes. Still no whale. We sit and stare and sweat and wait . . .

THE NOTION OF ARRANGING a peaceful encounter with whales is a bit ironic, of course, given the way humankind has treated them for centuries.

According to legend, in 1712, an American ship captained by Christopher Hussey was hunting right whales off the southern coast of Nantucket Island when a gale suddenly blew the vessel dozens of miles south, beyond sight of land, to a barren stretch of deep water in the middle of the Atlantic Ocean. The crew struggled to regain control of the ship and were readying the mast to tack back to shore when they noticed columns of mist shooting up at odd angles from the water's surface. Then they heard heavy, heaving exhalations. They had floated into a pod of whales. Hussey ordered the men to draw lances and harpoons and stab the whale closest to the ship. They killed it, tied it to the side of the boat, fitted the mast, and sailed back to Nantucket, then dropped the whale's body on a south-facing beach.

This was no right whale. Hussey knew that the mouths of right whales are filled with baleen, a hairlike substance used in filtering out krill and small fish. The whale he had just caught had enormous teeth, several inches long, and a single nostril on top of its head. The bones of its flippers looked eerily like those of a human hand. Hussey and his crew cut open the whale's head, and hundreds of gallons of thick, straw-colored oil oozed out. The oil must be sperm, they thought (wrongly); this strange whale must be carrying its "seed" within its oversize head. Hussey named it spermaceti (Greek *sperma*, "seed"; Latin *cetus*, "whale"). The English version of the name took hold: sperm whale.

From that point forward, the sperm whale was screwed.

By the mid-1700s, whale ships had flocked to Nantucket to join a thriving industry. Sperm whale oil, the straw-colored substance taken from the whale's head, turned out to be an efficient and clean-burning fuel for everything from streetlamps to lighthouses. In its congealed form, it made top-quality candles, cosmetics, machine lubricants, and waterproofing agents. The Revolutionary War was fueled by sperm whale oil.

By the 1830s, more than 350 ships and 10,000 sailors were hunting sperm whales. Twenty years later, those numbers would double. Nantucket was processing more than five thousand sperm whale corpses a year and reaping upwards of twelve million gallons of oil. (A single whale could yield five hundred or more gallons of spermaceti; oil from boiled blubber could produce about twice that amount.)

But hunting the world's largest predator didn't come without dangers.

Whalers in the eighteenth and nineteenth centuries were attacked regularly. The most famous incident occurred in 1820. The Nantucket whale ship *Essex* was off the coast of South America, its crew hunting whales, when they were rammed twice by a charging bull. The ship was lost. A crew of twenty men escaped in smaller boats and drifted off into the open ocean.

Nine weeks later, still drifting, the crew was close to starvation. Following maritime custom, the men drew lots to see who would be eaten. The captain's cousin, a seventeen-year-old named Owen Coffin, was chosen. Coffin put his head on the side of the boat; another man pulled the trigger of a gun. "He was soon dispatched," wrote the captain, "and nothing of him left."

Ninety-five days later, the boat was rescued. There were two survivors: the captain and the man who had pulled the trigger. The harrowing tale served as the basis for Herman Melville's novel *Moby-Dick* and, more recently, Nathaniel Philbrick's non-fiction bestseller *In the Heart of the Sea.*

AS SPERM WHALE STOCKS DECREASED in the ocean near Nantucket, and whalers had to search farther away, the cost of oil increased. Meanwhile, a Canadian geologist named Abraham Gesner invented a method of distilling kerosene from petroleum. This process produced a substance close to whale oil in quality, but much cheaper. In the 1860s, the whale-oil industry collapsed.

The discovery of petroleum sounds like a death knell for whaling, but ultimately, this new cheap fuel would hasten the sperm whale's destruction.

In the 1920s, new diesel-powered ships could proccss whale bodies so quickly and easily that whaling became profitable again. Sperm whale oil became a primary ingredient in brake fluid, glue, and lubricants. It was used to make soap, margarine, and lipstick and other cosmetics. The whale's muscles and guts were mashed up and processed into pet food and tennis-racket strings. (If you own a top-quality wooden tennis racket made between 1950 and 1970, chances are it was strung with the sinew of sperm whales.)

Whaling went global. From the 1930s to 1980s, Japan alone killed 260,000 sperm whales — about 20 percent of the total population.

By the early 1970s, an estimated 60 percent of the ocean's sperm whale population had been hunted, and the species was nearing extinction. While the world had grown proficient at hunting sperm whales, the whales themselves were a complete mystery. No one knew how they communicated or socialized; no one even knew what they ate. They had never been filmed underwater.

The documentary *Whales Weep Not,* which was seen by millions of people in the 1980s, offered the public the first view of sperm whales in their natural habitat. Sperm whales seemed far from the image handed down by history and literature. They were not surly brutes munching boats and men but gentle, friendly, even welcoming. The global antiwhaling movement gained support throughout the early 1980s and eventually ended all commercial whaling by 1986.*

The general increase in awareness of the sperm whale's intelligence and human-like behavior has not deterred some countries from trying to hunt them again. As of 2010, Japan, Iceland, and Norway have been pressuring the International Whaling Commission to end its thirty-year moratorium on whaling. Schnöller and other researchers predict the moratorium could be lifted as soon as 2016, and hunting of sperm whales could again become legal.

Sperm whales have the lowest reproductive rate of any mammal; females give birth to a single calf once every four to six years. The current sperm whale population is estimated at about 360,000, down from approximately 1.2 million just two hundred years ago, where it probably hovered for tens of thousands of years before whaling began. Nobody knows for sure, but many researchers fear the population has been declining once more. Continued hunting could significantly decrease the population for generations and eventually push sperm whales back toward extinction.

IF HUNTERS DON'T ERADICATE SPERM whales, pollution might. Since the 1920s, PCBs (polychlorinated biphenyls), carcinogenic chemicals used in the manufacture of electronics, have slowly seeped into the world's oceans and, in some areas, reached toxic levels. For an animal to be processed as food, it must contain less than 2 parts per million of PCBs. Any animal that contains 50 ppm of PCBs must, by law, be considered toxic waste and be disposed of in an appropriate facility.

Dr. Roger Payne, an ocean conservationist, analyzed sea life for PCBs and found that orcas had about 400 parts per million of

* Although some countries, like Japan and Korea, continued whaling under the loophole of "scientific research."

PCBs — eight times the toxic limit. He found beluga whales with 3,200 ppm of PCBs, and bottle-nosed dolphins with 6,800 ppm. All of these animals were, according to Payne, "mobile Superfund sites." Nobody knows how much more pollution (PCBs, mercury, and other chemicals) whales and other oceanic animals can absorb before they start dying off en masse.

Payne and other researchers point to the baiji dolphin, a freshwater native of China's Yangtze River, as a possible portent of the sperm whale's fate. Considered one of the most intelligent of all dolphin species, the baiji dolphin has become functionally extinct due to pollution and other manmade disturbances. (At last count, there were about three baiji dolphins left.)

For Schnöller and his colleagues, cetacean research feels like a race against time.

BACK ON THE BOAT, ANOTHER hour passes. And another. I check the thermometer on my dive watch and notice the temperature has climbed to 109.

Then, suddenly, an electronic chirp blasts from the back of the boat. It's Schnöller, calling our captain's cell phone. The DareWin team has just spotted a pod of sperm whales near the Trincomalee harbor. Schnöller says the whales have probably been there the whole time; we just hadn't been far enough out to spot them. They're following slowly behind the pod, waiting for an opportunity to get in.

The captain starts the motor and we shoot south. Soon we're surrounded by sperm whales.

"You see the ploofs?" says Prinsloo, pointing east at the horizon. What look like little mushroom clouds shoot from the surface at a 45-degree angle. A sperm whale has only one external nostril, which is located on the left side of its head and causes its exhales to emerge at an angle. These distinctive blows can go about twelve feet high, and on a windless and clear day, they're visible for a mile or more.

"They look like dandelions, don't they!" says Prinsloo. Three hundred yards to our right, another blow erupts.

"Get your mask," she says. "Let's go in."

Our team has agreed to put only two people in the water at any one time, to avoid scaring off the whales. I'm on the first shift. The captain turns and pulls parallel to the pod so that we're a few hundred feet in front of them.

"You can never chase down a whale," Prinsloo explains as she yanks off the sheet and grabs her fins. "They always need to choose to come to you." If we move slowly in predictable motions, just in front of the whales' path, they can easily echolocate the boat and get comfortable with our presence. If they're disturbed by us, they'll take a deep breath and disappear beneath the surface. We'll never see them again.

As the boat edges closer, the whales still haven't dived — a good sign. Prinsloo says it's not a full pod, just a mother and calf. Another good sign. Calves get curious around freedivers, and their mothers, in Prinsloo's experience, encourage them to investigate.

Both whales are four hundred feet from the boat when they slow down, almost to a stop. Our captain cuts the motor. Prinsloo nods to me; I pull on my fins, mask, and snorkel, and we quietly submerge.

"Take my hand," she says. "Now, follow." Breathing through our snorkels with our faces just below the surface, we kick out toward the whales. Today, the visibility is mediocre, about a hundred feet. We can't see the whales in the water, but we can certainly hear them. The blows grow louder and louder. Then the clicking begins; it sounds similar to a playing card stuck in the moving spokes of a bicycle. The water starts vibrating.

Prinsloo tugs my arm, trying to get me to hurry up. She pops her head above the surface for a moment and stops. I stick my head up and see a mound a hundred feet in front of us, like a black sun rising on the horizon. The clicking grows louder. The mound pops up from the surface again, then disappears. The whales leave; we don't see them depart. But we can hear them beneath the water, their blows softening as they drift off. The waters calm, the clicks slow like a clock winding down. And they're gone.

Prinsloo lifts her head and faces me. "Whale," she says. I nod, smiling, take the snorkel out of my mouth, and begin to tell her how incredible the experience was. Then she shakes her head and points behind me.

"No. Whale."

The mother and calf have returned. They're stopped and are facing us in the other direction, a hundred and fifty feet away. The clicking starts again. It's louder than it was before. I instinctively kick toward the whales, but Prinsloo grabs my hand.

"Don't swim, don't move," she whispers. "They're watching us."

The clicks now sound like jackhammers on pavement. These are echolocation clicks; the whales are scanning us inside and out. We watch from the surface as they exhale. With a kick of their flukes, they lunge toward us.

"Listen," Prinsloo says urgently. She grabs me by the shoulder and looks directly at me. "You need to set the right intention now. They can sense your intent." I know how dangerous human-whale interactions can be, but I strive to set my fear aside, calm myself, and think good thoughts.

Behind Prinsloo, the whales approach, hissing and blowing steam—two locomotives. "Trust this moment," she says. The whales are a hundred feet, seventy-five feet away. Prinsloo grips my hand. "Trust this moment," she repeats, and she pulls me a few feet beneath the surface.

A hazy black mass materializes in the distance, growing larger and darker. Details emerge. A fin. A gaping mouth. A patch of white. An eye, sunk low on a knotted head, peers in our direction. The mother is the size of a school bus; her calf, a short bus. They look like landmasses, submerged islands. Prinsloo squeezes my hand and I squeeze back.

The whales come at us head-on. Then, thirty feet from colliding, they pull softly to the side and languorously veer left. The rhythm of the clicks shifts; the water fills with what sounds to me like coda clicks. I believe they are identifying themselves to us. The calf floats just in front of its mother, bobs its head slightly, and stares with an unblinking eye. Its mouth is turned up at the end, like it's smiling. The mother wears the same expression; all sperm whales do.

They keep their gaze upon us as they pass within a dozen feet of our faces, shower us with clicks, then retreat slowly back into the shadows. The coda clicks turn to echolocation clicks, then the echolocation fades, and the ocean, once again, falls silent.

I RECALL FRED BUYLE TELLING me about a friend of his who had been freediving in the Azores, the Portuguese islands off the west coast of Africa, when a pod of female sperm whales approached. The whales showered him with clicks and gently interacted with him over the course of many hours. Then a young bull approached and, according to Buyle, "probably got pretty jealous." The bull turned toward Buyle's friend and shot him with clicks powerful enough to stun him. He managed to kick to the surface and crawled back to the deck of the boat, where he experienced debilitating pain in his stomach and chest. After three hours, he recovered fully and has since suffered no ill effects.

Schnöller told me a similar story. He was diving with sperm whales in 2011 when a curious calf approached and started bumping him with its nose. Schnöller held out his hand to push the calf back and felt a sudden shock of heat rush up his arm. The energy from the clicks coming out of the calf's nose was strong enough to paralyze Schnöller's hand for the next few hours. He too recovered.

Prinsloo and Ghislain had their share of close calls in Trincomalee last year. After spending hours in the water with a pod, a bull approached Ghislain at a fast clip. Prinsloo motioned to Ghislain to get out of the way. Just then, the bull turned, raised his twelve-foot fluke above the surface, spun it, and slapped it down. If Ghislain hadn't moved, his head would have been crushed.

Prinsloo and Ghislain claimed the fluke slap was possibly a playful interaction, not meant to harm. But when you're in the water with an animal five hundred times your weight and ten times your size, such play can be fatal.

Freediving was a precaution against accidents like this and an essential skill when interacting with whales. Whales, especially calves, get excited during human-whale encounters and can sometimes charge and smother divers. Being able to dive down to forty feet and stay there for a minute or so until a whale passes could mean the difference between a life-changing and a life-ending experience. The fact is that nobody – not Prinsloo, Schnöller, or Buyle – really knows how risky these kinds of encounters are. Up until ten years ago, Schnöller told me earlier, nobody was diving with whales.

"Everyone thought it was too dangerous," he said. Today, only a handful of divers attempt it, and most have had their share of narrow escapes.

Universities and oceanographic institutions would never allow their researchers or students to dive with whales. Few would want to.

Luke Rendell, a sperm whale researcher at the University of St. Andrews in Scotland, told me in an e-mail that Schnöller's research approach looked like "a pile of hokum" and was probably a "pretty flimsy scientific cover to go swimming with whales." He finished by saying, "I'm perfectly capable of collecting data without freediving with the animals, thank you." To be fair, Rendell said he welcomed more researchers to the field, but he thought the DareWin's website looked suspiciously like pseudoscience.

Schnöller brushes off the criticism as "normal scientist reaction" and insists that most researchers don't understand his work and those that do dismiss it as unscientific. But nobody can deny that Schnöller's freediving approach gets results.

In six years of freelance work, he has collected more video and audio data on sperm whale interaction and gotten closer to these animals than most institutional scientists do in decades. Institutional scientists study sperm whale communication by recording clicks with a hydrophone from the deck of the boat, without ever knowing which whale is clicking and why. One of the longest-running sperm whale research programs is the Dominica Sperm Whale Project, headed by Hal Whitehead. The group studies sperm whale behavior by, among other things, following pods around and snapping photographs of flukes when the whales come up for air.

Meanwhile, last year Schnöller had a face-to-face encounter with five sperm whales that lasted three hours. The entire dive was documented in three-dimensional video and high-definition audio and is, to date, the longest and most detailed footage of sperm whales ever recorded.

Schnöller has since made inroads with the French scientific community and is working with renowned cognitive scientist Fabienne Delfour and acoustic scientist Diderot Mauuary on DareWin's first scientific paper. They hope to publish it with Stan Kuczaj in a peer-reviewed journal in the next year. "This will all be official, it will all be scientific," Schnöller insists. He says he's not trying to subvert the scientific system — he wants to work within it — he is simply trying to speed up the collection of data, which, at the institutional level, happens at a glacial pace. For Schnöller, and perhaps for the whales, that pace may be too slow.

Sitting at the patio table that evening in the afterglow of my first sperm whale dive, I'm beginning to feel a bit of his frustration.

In the short encounter I had with Prinsloo, the mother and calf didn't make a U-turn to escape us. They came back to say hello. My eye-to-eye interaction with these animals was one of the most powerful experiences I've ever had. I felt a sudden recognition, an instant and ineffable sense of *knowing* that I was in the presence of something extraordinarily powerful and intelligent. This is not a scientific observation, of course, but an emotional one. Still, I believe it is as true and telling as any objective fact we may discover about these animals. You can't get that by sitting on the deck of a boat and sticking a robot in the water. You need to get wet.

ON THE FOURTH DAY, the film crew leaves. The cameraman who had been violently seasick since the first day has refused to spend another ten hours motoring around in a rickety boat. The director, Emmanuel Vaughan-Lee, was exhausted, and he and Schnöller weren't getting along.

"You never told me it would be this hard," said Vaughan-Lee when I'd talked with him that morning. He was scratching his bald, sunburned knees beside the patio table. I had warned him, repeatedly, but the point was moot. He told me he'd decided to cut his losses and take the next flight home to San Francisco.

He left a day too soon.

The remaining team of seven, plus the hired boat hands, cram into a single boat designed for half our number. With the motor coughing and wheezing, we head south. Hours later, we're fifteen miles from the coast and idling over the deep water of Trincomalee Canyon again. Schnöller checks his GPS, putting us near where they saw the whales yesterday.

"Turn the motor off. I listen for them," he says. From the bow of the boat, he grabs a sawed-off broomstick with a metal pasta strainer tied to the end. He inserts a small hydrophone into the center of the strainer and drops the whole contraption in water, then puts on a pair of ratty headphones.

This strange device, which is wired to an amplifier, works like an antenna to home in on sperm whale clicks. By spinning the strainer around, Schnöller can determine what direction they're coming from. Frequency and volume give him an idea of how deep the whales might be.

"They sell these to institutions for fifteen hundred euro," he says, laughing. "I make mine from junk, and it works just as good." Click Research, a new oceanographic manufacturing company he's now building, will offer a version that works as well as the institutional model for only $350.

Schnöller puts the headphones over my head and hands me the broomstick. "What do you hear?" he asks. I tell him I hear static. Schnöller cups the headphones tightly over my ears. "Now listen. *What do you hear?*"

He takes the broomstick from my hands and spins the strainer slowly around below the surface. Through the static, I begin to hear a syncopated rhythm, like distant tribal drums. I tell Schnöller to stop moving the strainer. Everyone on the boat falls silent. The rhythm speeds up and grows higher in pitch, the patterns overlapping. What I'm hearing isn't drums, of course, but the echolocation clicks of sperm whales hunting in the canyon miles beneath our boat.

Schnöller grabs the headphones and passes them around the boat. Everyone is entranced. A boat hand listens for a moment, then passes the headphones back to Schnöller. He gingerly walks to the bow and picks up a worn, wooden oar, then dips the paddle in the water and sticks the end of the handle in his ear.

The boat hand explains in stilted English that this was how Sri Lankan fishermen used to listen for whales hundreds of years ago. Sperm whale echolocation, even from miles beneath the ocean's surface, is strong enough to vibrate five feet of wood and make an audible clicking sound. I give it a try and hear a faint *tick-tick-tick*. It sounds like a signal from another world, which, in a way, is precisely what it is. I get chills listening to it.

Schnöller puts the headphones on and spins the strainer dexterously. He tells us the whales will switch from making echolocation clicks to codas as they ascend. By listening to these subtle shifts in click patterns, and the volume and clarity of the clicks, he has taught himself to predict the location and moment that the whales will surface, with startling accuracy. I ask him: How accurate? Then he demonstrates.

"They are two kilometers that way," he says, pointing west. "They are coming up. They will be here in two minutes." We sit, staring westward. "Thirty seconds . . ." he says. "They are moving to the east, and . . . right . . ."

Exactly on cue, a pod of five whales surfaces about fifteen hundred feet from our boat, each exhaling a magnificent blow. He grins, obviously proud of himself, takes off the headphones, and throws the strainer and broomstick in the bow. I give him a high-five. The boat captain looks dumbfounded.

"Okay, now," says Schnöller. "Who wants to go in?"

• • •

AFTER DINNER, SCHNÖLLER, Gazzo, and Ghislain are sitting around the patio table going over the day's footage. The clips are hypnotizing. Each of us had short encounters with half a dozen different whales. Schnöller and Gazzo recorded the interactions in 3-D high-definition video. He says this is the first time some of these behaviors have been documented at such close range. The most impressive footage, he says, comes from the dive that Guy Gazzo and I took at the beginning of the day.

A pod of about five whales turned and approached our boat. Schnöller told me to grab my mask and follow Gazzo, who was carrying the 3-D camera, into the water. At first the whales were moving away from the boat, but as we swam out farther they changed direction to meet us face to face. Some two hundred feet in front of us, a shadow expanded, then separated into two forms — two enormous whales, perhaps thirty-five-feet long. One whale, a bull, came directly at us but then unexpectedly spun around so that its belly was facing us. We couldn't see its eyes or the top of its head. As it approached, it dove just beneath our fins and let out a rapid burst of coda clicks so powerful that I could *feel* them in my chest and skull. The bull, still upside down, released a plume of black feces, like a smoke screen, and disappeared. The entire encounter lasted less than thirty seconds.

Schnöller boots up the video on his laptop and plays it back for me. This time, he turns up the volume on his speakers.

"You hear that?" he says, then reverses the video again, and again. I listen closely. The clicks sound harsh and violent, like machine-gun fire. "That's not a coda." Schnöller laughs. He plays the clicks back again. "And he's not talking to you."

What Gazzo and I heard and felt was a creak — the echolocation click train that sperm whales use when they're homing in on prey. The whale flipped on its back so it could process the echolocation clicks more easily in its upper jaw, much as a human might cock his head to focus on a sound. Schnöller plays the video again and again, laughing.

"He was looking at you to see if he could eat you!" he says. "Lucky for you, I guess you didn't look too delicious."

But this brings up a question I've had ever since we first boarded the boats. Why didn't they eat us? We're certainly easy prey.

Schnöller believes that, when the whales echolocate our bodies, they perceive that we have hair, big lungs, a large brain — a combination of characteristics they don't see in the ocean. Perhaps they recognize that we're fellow mammals, that we have the potential for intelligence. If this theory is correct, then sperm whales are smarter than us in one crucial way: they see the similarities between our two species more readily than we do.

He then brings up another file on his computer, a ten-second audio loop he recorded with the hydrophones earlier in the day. He clicks Play.

"Well?" He looks at me. I tell him the only thing I hear is distant echolocation clicks, which sound like random emanations from a drum machine. He orders me to put on his headphones, turns up the volume, and blasts me with what sounds like an enormous bomb exploding from miles away.

"I think this is something big," he says. I ask him if the hydrophone just bumped into the side of the boat. "No, impossible," he says. "This is something important. I promise you, this is big."

IN ORDER TO EASE THE constant bickering between Prinsloo and Schnöller, we decide to rent another motorboat for the next several days. Schnöller's team will take one boat while Prinsloo takes the other. I'll alternate between the two, starting with Prinsloo.

During the off-hours when there are no dolphins or whales to dive with, we cut the motor, jump overboard, and freedive. Prinsloo brings her training float and rope and suggests we do some freediving depth training.

"How deep do you want to go?" she asks me, sitting cross-legged at the prow. "Fifty feet?" Without waiting for my reply, she grabs her mask and fins, jumps overboard, and swims the float out a dozen feet. I put on my gear and follow. The water conditions today are perfectly clear, with visibility extending perhaps two hundred feet. Even farther below, the depths aren't black and brooding like they had been at the 40 Fathom Grotto, but a violet-blue. I watch through my mask as Prinsloo ties a weight belt to the end of the rope and releases it down into the water. From where I'm floating, it looks like a root from a tree growing in time-lapse images.

Marshall suits up and swims near Prinsloo; the two breathe up together, then dive down in unison along the rope. I begin the pre-dive breathing exhalations — "Inhale one, hold two, exhale ten, hold two" — and close my eyes, trying to calm my thoughts and relax my body. I focus on the static breath-holds I'd been doing for the past few months, and I try to remember how easy it was to hold my breath for three minutes, and how easy it will be to hold my breath for just one minute as I dive down fifty feet to the end of the rope and return.

It's a lot of self-talk, but all my coaches have told me this internal cheerleading is essential: I must convince myself this dive will be easy *and* enjoyable. Freediving, as William Trubridge said, is a mental game.

When I open my eyes a few minutes later, I'm a little light-headed from all the heavy breathing, and I feel a strong sense of vertigo. Prinsloo and Marshall, whose tiny figures swim in circles at the end of the rope, appear to be floating high in a cloudless sky. The scene seems in every way a mirror image of the surface world; there are no other markers — underwater animals, seafloor, boat bottoms — to convince me otherwise. Luckily, after months of training, I'm getting used to this kind of disorientation. I relax and roll with it.

*Inhale one, hold two, exhale ten, hold two . . .*

I begin my final ten exhalations. My mind returns to the training with Prinsloo in Cape Town. We were at a freshwater quarry with four other students, practicing deep dives. I was, again, having difficulty making it below twenty feet. I resurfaced from a painful attempt to thirty feet when Prinsloo approached me from the other side of the float and suggested I try the next dive, the deepest of the day, with my eyes closed. This was an exercise in trust, she said; I needed to trust her, and I needed to trust myself. I thought it was a horrible idea, but I didn't say that. I didn't say anything. I inhaled, squinted, dove down, and resurfaced one minute later having completed the deepest, longest, and most comfortable dive I'd ever made. I dove to a depth of forty feet without feeling the slightest discomfort.

NOW, STARING DOWN INTO THE void of blue water, watching Prinsloo and Marshall hovering six stories below me, I try to remember how that blind dive felt. Then I take one last inhale and descend.

As I pull with my left hand, I reach down and pinch my nose with the right, lift a puff of air from my stomach into my head, then cough a *T* sound into my closed mouth and seal my throat with my epiglottis. I jackhammer that trapped air from the back of my throat up into my sinuses. It's the first time I've used the Frenzel method at depth. It works seamlessly. After half a dozen pulls, I've passed twenty feet and am falling quickly.

The pulling gets easier the deeper I go. I'm able to loosen my grip so that I'm pulling the rope with just the thumb and forefinger of each hand. Moments later, I let go entirely. I'm neither kicking nor pulling, but I continue to lunge downward. I've hit neutral buoyancy — zero gravity. The door is open. I bring my arms down to my sides in skydiver pose and prepare to fall into deeper water.

First, the wetsuit vest tightens. My chest feels like it's been shrink-wrapped. My lungs push up toward my throat, and my stomach caves in slightly. This is the pressure of the deep ocean pressing on my exterior; it's inside me too, sucking my body into itself like a black hole.

The huge breath I took at the surface has disappeared. I didn't exhale it; I've been holding it the whole time. But it's now compressed to half its volume, enough so that it tugs at the soft tissues of my lungs and throat. This sounds uncomfortable, but it isn't. The feeling is unexpectedly warming, as if someone just threw a blanket around me. It's the feeling of peripheral vasoconstriction, of oxygenated blood flooding in from my arms, legs, hands, and feet into my core.

I've just flipped the Master Switch.

Months back, when I was in Greece, I asked a competitive diver what it was like to dive deep, to have a hundred pounds of pressure per square inch pressing against her body. Her answer then struck me as woo-woo: she said it felt like the ocean was hugging her. But that's exactly how it feels, a generous squeeze from the largest mass on the planet.

I drift farther down. I feel a buildup of pressure in my ears, and it's more painful than any I've experienced before. I pinch my nose and try to equalize but can't; the air that I had trapped in my sinus cavities, like the rest of my body, has been halved in volume. My lungs feel completely empty too, but I know from my training with Ted Harty that this is an illusion. There's still plenty of air to draw from.

Though I've let go of the rope, I've never strayed from it, and now I grab it again with my left hand to stop my descent, then back myself up a few feet to let the air re-expand in my sinuses and ease the ear pain. I pinch my nose again and draw another puff of air from my lungs into my head, then shuffle it between my nose and ears. My ears open with an audible squeak, then a *pop*. I let go of the rope with my left hand, kick a fin to restart the descent, and drop deeper.

The weight belt at the end of the rope passes by the length of my body — my chest, legs, feet, then fins. I'm descending at about the same rate that a feather drops in the air. In front of me, there's no longer any rope. All directions glow in the same neon-blue light and it goes on forever.

Part of me wants to keep going, to continue exploring this alien space. I feel no onset of convulsions, no nagging need to breathe, no coldness, not even a strong sense of being underwater. But I know this is competitive freediving's temptation speaking to me — *Go deeper,* it says. And that's not the kind of freediving I've come here to master.

I grab my knees, curl into a ball, and flick my right fin to turn my body around in a slow-motion somersault. The world flips upside down, and the vertigo I experienced at the surface restarts.

Now it appears as if I'm no longer floating at the end of the rope but hovering from above, preparing to fall to earth. I kick a few feet up, grab the weight belt with my right hand, and dangle there for a moment.

The first few pulls require some effort; the force of 200,000 pounds of water squeezes against me, trying to tug me farther down. A few big pulls, a few strong kicks, and I'm back in that zero-gravity zone. The pulling here is significantly easier. The air that vanished from my lungs and head at depth now miraculously returns. It feels like my chest is being inflated with a pump. There's no need to equalize my ears on the ascent; the expanding air inside my head does that automatically. And I can ascend as quickly as I want. My body, like all human and most mammalian bodies, is specially adapted to process the exchange of oxygen and nitrogen gases that occur during a deep dive – a trigger of the Master Switch.

I pull on the rope now with both hands and kick my fins with more force. The same invisible hand that drew me to the depths is now lifting me back up to the surface. I'm ascending at twice the speed I descended. I've reached the positive-buoyancy zone.

Nearing the top of the rope, I look skyward and see the reflective sheen of surface water; the float and boat bottom are now less than twenty feet away. The air within my lungs expands by another third. It feels like a living thing trying to get out. I open my mouth, relax the epiglottis in my throat, and a cloud of bubbles and vapor streams from my mouth. Seconds later, my head breaks through to the surface atmosphere and I'm spitting out flecks of water, breathing in fresh air, and blinking my eyes in the flash-bulb-bright morning light.

There's no sense of flushing in my face; no quivering in my stomach; no need to gasp for air; no aching ears, throbbing headache, or dizzying highs. There is no pain at all.

Marshall and Prinsloo are floating a few feet beside me. Prinsloo's been watching the whole dive. She doesn't say anything; she doesn't congratulate me or ask how deep I've dived. She doesn't even acknowledge that she's been monitoring me. There's no boasting here or judges to impress. This is no competition.

Without saying a word, all three of us breathe up, upturn our bodies, then, all together, fall back down past the doorway to the deep.

# −35,850

IT TAKES A LONG TIME to become ooze. First you need to die and be eaten, then excreted, then have another organism eat that excrement, then have yet another animal eat that organism that just ate that excrement, and so on. This cycle will repeat until all that's left of you are a few million molecules spread out like a constellation of stars across the world's oceans. And you've still got a few thousand years to go before you become ooze.

At some point along the way, one of those tiny bits of you will leave the food cycle and be pulled down to deeper water. During this descent, you'll be surrounded by phytoplankton that will degrade you into even smaller bits. When these phytoplankton die off after a few days, the last little bit of whatever is left of you — some cluster of molecules — will drift off inside the microscopic skeletons. These will join trillions of other tiny skeletons in a never-ending snowstorm of detritus that floats down to deeper water.

Most of these particles will be recycled by the time they reach ten thousand feet. Only a fraction of 1 percent will make it to the seafloor below twenty thousand feet, a depth so dark and fore-boding that scientists have named it the hadal zone, from the Greek word *Hades,* or "hell."

Now comes the hard part. To become ooze, these last tiny bits of you must sit on the bottom of the deep sea, undisturbed, and solidify for hundreds, thousands, even millions of years.

This ooze of microscopic skeletons blankets more than half the ocean floor. Billions of years ago, when the ocean covered the planet, ooze coated what is now land. Look around and you'll see its remnants everywhere. The pyramids of Giza were built using limestone, a sedimentary rock made up of ooze. London's Houses of Parliament and the Empire State Building are also built us-ing limestone. The concrete sidewalk in front of your house is filled with ooze. You probably brushed your teeth with ooze this morning. (That white stuff in toothpaste is made from calcium carbonate, a chalky compound composed partly of ancient phy-toplankton skeletons.) The silicon in computer chips that power the e-readers on which some of you are now reading these words come from the same siliceous microscopic shells that oozed on the seafloor millions of years ago. Our world is built on micro-scopic bones.

Doug Bartlett, a lanky man with round glasses and kind eyes, knows all about ooze, and just now he's pointing some out to me, in a seawater sample that he keeps inside a pressurized stainless-steel tube. Bartlett is a marine microbial genetics researcher at the Scripps Institution of Oceanography in La Jolla, California. He has studied ooze for the past twenty-five years, and he's spent a decade collecting it.

We're standing in a refrigerated room just down the hall from his office, a storage locker containing dozens of tubes that hold water samples from the world's deepest oceans. Inside these tubes are phytoplankton and microbes that will one day become ooze. Each tube is kept at the pressure of the depth from which it was taken, in some cases fifteen thousand pounds per square inch. This allows Bartlett and his team to study the microbes in their original form, in their natural environment, and, in some cases, to grow microbe cultures, a substance akin to deep-sea yogurt.

"We're like astronomers looking up at the heavens at millions of stars," says Bartlett. "Instead of telescopes, we use microscopes and see billions of microbial life forms." He tells me that there is more diversity among microbes than any other life form, and that the deep ocean harbors the greatest diversity of microbes of any environment on the planet. By studying these microbes, Bartlett and his team hope to figure out how the Earth might have formed billions of years ago, where life on the planet first began, and, possibly, where all life might be heading one day.

These are hard questions, made harder by the location of the answers: 20,000 to 35,850 feet down. That's the depth of the hadal zone, the world's deepest ocean region and where Bartlett gathers his most valuable samples. To get there, Bartlett and his team have built unmanned robots, called landers, that they send down to the seafloor to suck water into pressurized vaults, gather samples, and, perhaps, capture some never-before-seen animals. Sinking a lander in the deep water is relatively easy. You just drop it, and gravity does the rest. Retrieving it is another story. Unlike ROVs, bathyspheres, and other deep-sea research vessels, landers aren't tethered to support ships, and they don't have motors. Instead, they use an almost-quaint system of weights and air packets.

The lander that Bartlett's engineers are preparing outside his office for an upcoming expedition has a fifty-pound platform of dead weights strapped to the bottom, which will pull it down to the seafloor. When the lander hits the ocean bottom, the engineers will send an acoustic signal from a deck box (basically, a sonar device) that will command it to suck water into a pressurized canister. Bartlett won't know where the lander hit the seafloor or what it might be sampling until he gets it back aboard his boat. These robots drive blind.

Roughly an hour after sampling, the engineers will signal the lander to release the weight platform, causing it to float back to the surface. A radio transmitter mounted on the side will send out coordinates to a retrieval vessel. They'll motor over in the general location of the coordinates, keeping a lookout on the water below. Flashing beacons will help them locate the lander if it comes up at night.

That's how it's supposed to work, anyway. Lander research is a relatively new science and only a half a dozen researchers do this kind of work in the hadal zone. Things go wrong all the time. In the decade that Bartlett has been collecting samples, he's had landers break, malfunction, go missing — sometimes all at once.

Hadal research is made even more difficult by the fact that most of the world's deepest oceans are located hundreds of miles from land, often off the coast of distant countries. Shipping a container filled with thousands of pounds of this stuff across an ocean to a dingy port in Guam or Mexico (something Bartlett has done numerous times) is a logistical nightmare and very expensive. Then there are the days at sea needed to reach deep water, a journey that can cost tens of thousands of dollars in fuel and boat-rental fees.

Putting all this together, you begin to understand why so few scientists — and absolutely no freelancers — do hadal research. No citizen-scientist has that kind of money. Most universities and research institutions don't either. Bartlett is one of the most established and respected deep-sea microbiologists in the world; he works at one of the most renowned oceanographic institutions. Still, he and his team manage to get out to research in the "laboratory" about once a year, if they're lucky.

I GOT TO SEE FIRSTHAND just how difficult it is to conduct hadal research. During a phone interview, Bartlett mentioned that he was interested in returning to Sirena Deep, a depression going down more than thirty-five thousand feet in the Mariana Trench, the deepest trench in the world. He suggested I come along. Then he suggested that I help organize the trip. We blocked out some dates in the summer and I started calling around.

The advantage of going to Sirena Deep, Bartlett said, is that it's located only ninety miles south of the North Pacific island of Guam, itself "only" a fifteen-hour flight from our homes in California. The disadvantage, I soon learned, is that Guam is a U.S. territory, which means all vessels harbored there must abide by U.S. maritime rules and regulations. This makes hiring a research ship nearly impossible. Within a week, I had run up a sizable long-distance phone bill. I talked to every harbormaster and yacht club in Guam looking for a vessel that could legally make the trip.

Most fishing boats large enough to handle the landers and our five-person crew didn't have the fuel capacity or accommodations; the few that did were forbidden by U.S. law to carry civilians more than twenty miles from the coast. Commercial ships, like tugboats, had the capacity and permission but were outrageously expensive. One captain quoted me a price of $80,000 for a two-day rental — about ten times what Bartlett's budget allowed.

And so it was quite a coup when, in October 2014, four months after the hardcover publication of this book, Bartlett made contact again. He and his team had finally secured a week-long expedition to the Mariana Trench, the deepest point in the world's oceans. And there was an extra cot on the boat.

Their plan: to deploy landers that could capture water samples, microbes, and other never-before-seen life forms. They'd also attempt high-definition sound recordings in the trench to understand how animals in this black and super-pressurized environment might use sound to hunt, mate, and perhaps communicate. It would be cutting-edge research happening a mile deeper than Mount Everest is tall.

Best of all, the expedition would take place aboard the R/V *Falkor*, a 272-foot German research vessel owned by the Schmidt Ocean Institute, an oceanographic research institute founded by Google CEO Eric Schmidt and his wife Wendy. The *Falkor* is not only among the world's most technologically advanced deep-sea research vessels, but also the most luxurious. Gone were the days of stale croissants and warm knockoff Fresca in rickety fishing boats, flea-infested child-size beds, and sun poisoning. I'd be traveling on the *Queen Mary 2* of research vessels, a floating Club Med for oceanographers.

Within a day of hearing from Bartlett, I purchased an airline ticket from Buenos Aires, where I was living for a month, to Hagåtña, Guam, some ten thousand miles northwest in the Pacific Ocean, where the R/V *Falkor* would disembark. It would be my last splash into the deep, this time to the lowest natural point on the planet.

IMAGINE YOU'RE LOOKING AT THE world's tallest mountains on a three-dimensional topographical map. You'll notice mountain ranges covering all the continents: Everest and K2 in the Himalayan range, Kilimanjaro on the east coast of Africa, Mont Blanc in the French Alps, Mount McKinley in the Alaska Range, and so on. Now imagine flipping that map upside down and looking at each of the tallest mountains from underneath. You'll see that the world's highest peaks have suddenly become the world's deepest trenches. This is what the ocean floor looks like, and those deep trenches dotting the seafloor are what scientists refer to as the hadal zone.

Hadal zone trenches, just like the world's highest mountains, are scattered across the globe, separated by hundreds, sometimes thousands, of miles. In other words, the hadal zone is not a contiguous stretch. What these discrete areas have in common, and what qualifies them as a single zone, is that they are all located between 20,000 and 35,850 feet down.

The pressure here ranges from 600 to 1,050 times greater than it is on the surface. If you could swim down there – which you couldn't – it would feel roughly the same as balancing the Eiffel Tower on top of your head. Then there's the temperature, which hovers just above freezing. There is no light, of course, and even oxygen is in short supply.

Although all the standard building blocks of life – sunshine, oxygen, heat – are missing from these waters, life somehow persists.

In 2011, Bartlett and a team of researchers dropped a lander rigged with lights and video cameras into Sirena Deep to a depth of about 35,000 feet. They hoped to see deep-water shrimp, maybe a few rocks, some ooze. What they discovered instead was a congregation of giant amoebas the size of a large man's fists rooted in the seafloor, each covered in a coat of frilly appendages that resembled the ruffles of a 1970s-era tuxedo shirt.

These creatures, called xenophyophores, were more than four inches wide, and yet each was a single cell. Xenophyophores have no brains or nervous systems, but they were able to mate and feed among the deposits of million-year-old detritus, seemingly unaffected by the fifteen thousand pounds of pressure per square inch that weighed on their bodies. To make the scene even more bizarre, halfway through the video, a jellyfish swam lazily by, the deepest ever filmed.

Bartlett had discovered the world's largest species of single-celled organisms living in the world's deepest ocean.

A year later, in 2012, a group of hadal researchers from the University of Aberdeen in Scotland sent a metal trap down 22,000 feet into the Kermadec Trench, the world's second-deepest, which is located off the coast of New Zealand. A few hours later, they pulled up an albino shrimp the size of a housecat.

Four years earlier, this same group had discovered schools of foot-long fish, called snailfish, at depths below 25,000 feet. The snailfish had fins shaped like bird wings, and instead of eyes, they used vibration sensors on their heads to find their way around.

Until very recently, scientists thought the hadal zone was a desert. The few animals that lived down there were supposed to be viscid, scrawny, small, and inactive, like the ones in shallower waters. But snailfish were fat and happy-looking, swimming briskly along the seafloor, interacting with one another as if they were members of the same family.

Nobody had expected such a profusion of life because nobody had looked; the technologies Bartlett and the team at Aberdeen were using were new, and each of these landers was specially designed for the mission.

At this writing, scientists have discovered at least seven hundred unique animal species in the hadal zone. An estimated 56 percent of these animals are endemic to hadal water, meaning they probably live nowhere else in the ocean. Further, only 3 percent of those endemic hadal animals were found in other hadal zones.

What these discoveries suggest is that each of the hadal zones pocking the ocean floor could have its own distinct life forms. And these life forms could have been on their own evolutionary path for millions of years.

It's as if there were an archipelago of Galápagos Islands buried beneath five miles of black ocean, removed from the rest of the world and developing new life in wondrous ways. And they've been there for millions of years, just waiting for us to shine a light on them.

This multiworld theory can be proved or disproved only through more extensive deep-sea research. Sadly, not many people are looking. Beyond Bartlett's team and the Aberdeen group, only a handful — Bartlett literally counted them on one hand — of other researchers in the world have the resources, and interest, to explore life below twenty thousand feet.

Today, the hadal zone remains one of the most poorly investigated habitats on the planet.

A MONTH AFTER I BUY my ticket, I'm in the Port of Guam dragging a suitcase up the gangway of the R/V *Falkor*. It's about 8:00 p.m., and the vessel is buzzing with activity: Guys in baggy jeans and white hardhats hustle along the aft deck. A tall man wearing a captain's hat and speaking with a German accent gives orders on a walkie-talkie. Small metal doors open and close like panels on a cuckoo clock.

A petite woman with long brown hair, a wide smile, and braces greets me at the back door. Her name is Adrianna and she'll be taking me on a brief tour of the ship. She walks me through a narrow door with rounded corners and up a labyrinth of companionways. On one of the hallway's white, steel walls is a world map with a few dozen red thumbtacks marking all the harbors, trenches, and oceans the *Falkor* has visited since it began running scientific missions in 2012: the deepest waters of the Caribbean Sea, hydrothermal vents off the coast of Papua New Guinea, two-hundred-year-old sunken ships near Greenland, unexplored deep-water reefs in Roatan, oil spills in the Gulf of Mexico, and submarine volcanos off the coast of Oregon.

The R/V *Falkor*'s mission, according to Wendy Schmidt, Eric's wife and Schmidt Ocean Institute's cofounder, is "to communicate about the science of the oceans to people so that they can care about it." To that end, the Schmidts have invested tens of millions of dollars of their own money to convert the *Falkor,* a former fishing protection vessel, into one of the most advanced independent deep-sea research ships in the world.

The *Falkor*'s board selects about a half-dozen teams each year to conduct oceanographic studies on the ship. Those lucky enough to win research time get a state-of-the-art vessel with twenty crew members at their disposal for up to forty days at a time—at zero cost. And because the R/V *Falkor* is privately operated and not associated with any institution or organization, research here can happen quickly, nimbly, and without much of the red tape of governmental or academic institutions. This is, in essence, a DIY effort akin to DareWin and Stanley Submarines, only on a much grander scale.

Adrianna leads me further down the hall to one of the *Falkor*'s three on-site laboratories. The first, called Dry Lab, operates as a control center for the ROVs usually deployed off the aft deck. The room looks like a Star Trek film set. At the back of the lab, twenty-seven monitors of varying sizes stare back at me. Above them a huge LED clock and LED banners flash and scroll out the temperature and time every few seconds. Desks are crammed with dozens of keyboards, joysticks, telephones, computer servers, and other assorted unidentifiable electronic gizmos. At the center of all this are four oversized command chairs. The only thing missing here is the crew. Turns out they are on the next level down, in the mess hall, finishing up dinner.

TWO GOURMET CHEFS ABOARD THE *Falkor* prepare three meals a day for crew. Tonight's menu includes leg of lamb au jus, quinoa salad, green salad, pasta salad, tomato soup, roasted sweet potatoes, mashed potatoes, asparagus, and white chocolate bread pudding for dessert, all served on warmed plates with a choice of red or white wine. I grab a glass of Pinot Noir and Adrianna walks me back to my berth.

My quarters are about as narrow as a New York City hotel room but twice as clean and with two porthole windows that look out just above the water's edge. "Enjoy your stay," says Adrianna, closing the door behind her.

I rest my wine glass on the desk, fall onto the memory-foam mattress, wrap myself in high-thread-count sheets, and start chuckling. I had always anticipated that the deeper I plumbed the ocean, the more battered and unsavory my transport and environs would become. The R/V *Falkor* has proven me wrong. Sure, I'll spend the next week trapped behind two feet of steel in the hull of an enormous ship, never swimming—or even touching—the deep waters I've come to research, but somehow, lying here in an air-conditioned room, buzzed off a leathery Sonoma Pinot, I can hardly complain.

THE DOUG BARTLETT OF THE 1970s, and one of the founders of deep-sea scientific research, was an Oregon State University marine geologist named Jack Corliss. In 1977, Corliss chartered a research vessel off the coast of Ecuador and steamed out two hundred miles to the Galápagos Trench. He began trawling the ocean floor looking for hydrothermal vents—underwater geysers that spew lava and superheated, chemical-rich water from the Earth's molten core. Corliss suspected hydrothermal vents existed, but he had never seen one. Nobody had. Corliss wanted to be the first, and he had a hunch the Galápagos Trench was a good place to start.

Hydrothermal vents aren't exactly easy to find. They're scattered along deep-sea mountain ranges formed by plate shifts. Together, these ranges extend more than forty-six thousand miles and may house numerous vents, sometimes close to one another, sometimes separated by hundreds of miles.

The morning of the first day above the trench, Corliss's crew lowered an ROV named *Angus* into the water and prepared for the first dive. As cables spooled out from the deck and the ROV sank deeper, Corliss stepped to the observation deck. He stared at a monitor as *Angus* plunged a thousand, two thousand, three thousand feet. At around eight thousand feet, the temperature gauge registered a huge spike — a good sign. Hot water that far down in the ocean meant that a hydrothermal vent might be close.

The engineers controlling *Angus* triggered the ROV's onboard camera to snap a series of photographs. They pulled *Angus* back on deck, removed the film from the underwater camera, and developed it. The grainy black-and-white photographs revealed not only the presence of active hydrothermal vents but crabs, mussels, and lobsters. There was life down there, tons of it, flourishing around a column of seawater hot enough to melt lead (750 degrees). The water didn't turn to steam like it would at the surface because of the tremendous pressure at these depths. They'd found a pressure cooker of life. Soon after, *Alvin,* a deep-diving sub from Woods Hole Oceanographic Institution, arrived on the scene. Two pilots hopped into the tiny submersible, plunged into the water, and followed *Angus*'s coordinates to the vents. Right on cue, at eight thousand feet, the temperature gauge rose. They looked out the port windows and motored cautiously toward an outcropping of steaming and smoldering white rocks.

"Isn't the deep ocean supposed to be like a desert?" one pilot said into a hydrophone connected to the support vessel above.

"Yes," a crew member replied.

"Well," said the pilot, "there's all these animals down here."

In front of Alvin were shrimplike creatures, albino crabs, mussels, lobsters, fish, anemones, and clams. Foot-long candy-cane-striped worms, an unknown species, swayed in the currents like wheat in a field. Corliss called the place the Garden of Eden.

Back onshore, scientists listened to the reports with extreme skepticism. And who could blame them? Until 1977, it was thought that all life required sunlight. Trees and plants need the sun's energy to convert carbon dioxide and water into fuel. Animals eat trees and plants. Even organisms that live deep underground or thousands of feet under the water and never see sunlight rely on the nutrients created by solar energy above. But not these animals. Corliss and his crew had stumbled upon not only a new species, but an entirely new biological system fueled by chemicals. Scientists called it chemosynthetic life.

The Garden of Eden would be recognized as one of the most significant scientific discoveries in human history.

THE REVELATIONS ABOUT CHEMOSYNTHETIC LIFE led to another mind-boggling discovery. It turned out that hydrothermal vents were only temporary homes. Vents die off and new vents suddenly erupt. Chemosynthetic life forms need chemicals and hot water to survive. While some animals, like shrimp, could alternate between photosynthetic and chemosynthetic environments, other hydrothermal animals, like mussels, could not. (Mussels hardly move, and they certainly couldn't travel a few hundred or a few thousand miles to the next deep-water trench to find another vent. They'd die en route.)

And yet, somehow, these mussels, crabs, tube worms, and other hydrothermal life kept showing up at each newly discovered vent community. Researchers estimate that hundreds of hydrothermal vents exist along the seafloor of the world's oceans, and most of them have never been seen. But even with the limited amount of exploration that has occurred to date, scientists have discovered six hundred new species of chemosynthetic life.

It isn't just the hadal zones that harbor their own unique animals and organisms; the vents do as well. The sea appears to be home to hundreds, perhaps thousands, of small, secluded biospheres.

The researchers found that the more inhospitable the environment, the more life seemed to flourish. The areas around vents, for instance, hosted up to a hundred thousand times more life than surrounding waters that weren't heated by vents. It was discovered much later that the deep pelagic realm, those waters from about 13,000 to 35,000 feet, housed the largest animal communities, the greatest number of individuals, and the broadest animal biodiversity of not only the ocean but any place on Earth.

Where did all this stuff come from?

Nobody knows for sure, but there's increasing evidence suggesting it came from inside the vents. Life on Earth may have started not on the sunlit surface but down in the boiling toxic water of the world's deep oceans.

ON OUR THIRD DAY AT sea, the *Falkor* arrives at Challenger Deep, a gaping hole at the southern end of the Mariana Trench and what scientists believe is the deepest point on the planet. Only three people have ventured down to these hellacious waters.

In 1960, Swiss engineer Jacques Piccard and U.S. Navy lieutenant Don Walsh took the plunge in a vessel called *Trieste*. The men crouched in a bathyscaphe, a steel sphere about the size of a phone booth, that was attached to the bottom of a sixty-foot-long tank of chambers filled with gasoline and iron pellets. (When the iron pellets were released on the seafloor, the gasoline, which is more buoyant than water, would float the vessel back to the surface.) A window broke on the way down, and Piccard and Walsh spent only twenty minutes on the seafloor, but they managed to make it back to the surface alive after an eight-hour dive. *Trieste*'s onboard depth gauge listed a depth of 37,799 feet (which was later corrected to 35,797 feet) — more than seven miles deep. A half century later, in March 2012, film director James Cameron took a submersible named *DeepSea Challenger* to a depth of 35,756 feet, to become the third person ever — and the only private citizen — to reach Challenger Deep's seafloor. Nobody's tried since, and it's likely nobody will try again anytime soon.

Scientists know precious little about life at these depths because so few researchers have ever explored these black and brooding waters. In the last twenty years, less than a dozen research expeditions have come to Challenger Deep, and still fewer

were able to do worthwhile research. The weather and sea conditions in this stretch of the ocean are often brutal; wind and swells
and tumultuous storms are the norm. Then there's the depths
themselves, where pressures can reach more than sixteen thousand pounds per square inch. Few ROVs can withstand such
stresses. For this reason, landers are the research tool of choice
for researchers, but even those often fail.

"It's a poker game, for sure," says David Barclay, an associate
professor at Dalhousie University in Halifax, Nova Scotia. Barclay, who is thirty-two, heavily bearded, and carries himself with
a Jack Black slackness, is leaning over the guardrail of the roof-
top above the navigation deck of the R/V *Falkor*. It's about 4:00
p.m. and Barclay has been up here for two hours, staring out at
the sea, hoping to see the flashing beacon of his lander somewhere on the horizon. Last night, Barclay launched Deep Sound
3, a lander rigged with an onboard computer and an array of hydrophones, hoping to capture the world's deepest audio samples.
Deep Sound 3 should have arrived back at the surface at 8:00 a.m.
today. That was eight hours ago. The beacon rigged to the top
of the lander was supposed to send GPS coordinates every few
minutes, but Barclay hasn't heard anything. "It's not looking too
good," he says, sipping a Sunkist. "Our best hope is to wait until
nightfall and hope we can spot the flashing beacon."

Doug Bartlett and his team aren't faring much better. Yesterday morning he deployed a lander named ARI, after Art Yayanos,
a retired deep-sea researcher from Scripps. ARI was rigged with
specialized samplers to capture water from Challenger Deep and
store the samples at the same ambient pressure and temperature
as the seafloor. Doing this would keep the bacteria within the
samplers alive, allowing Bartlett and his team a rare view into
how life can survive at hadal depths. ARI was supposed to resurface around midnight. Bartlett hasn't received a GPS signal
or seen any sign of the lander. Now, as we approach nightfall, it
appears as if our team has lost its two most important pieces of
equipment on the very first deployment, at a cost of about two
hundred thousand dollars.

I grab the rail on the starboard side opposite Barclay and, together, we silently scan the maw of ocean, which, today, is a violet-purple hue, the color of a day-old bruise. The water here not only looks different from any other ocean I've seen; it *feels* different too. This may be all in my mind, but I get a strange sense of vertigo knowing I'm hovering over such depths. It's the same feeling you get craning your head up at a skyscraper, but the exact opposite. Instead of light-headedness, there's a constant sensation of being pulled down, as if your body is covered in a heavy shroud. If the water below us were suddenly drained out, we'd fall the equivalent of twenty-five Empire State Buildings until we reached the bottom. If the seafloor were suddenly transposed above us, it would be in the stratosphere, the altitude at which long-haul airplanes fly.

These Challenger Deep analogies keep looping in my head — The Stratosphere of the Sea; The Mount Everest of Inner Space — until a minute becomes an hour, daylight turns to night. Barclay and I sigh and stare and wait.

IN THE 1980S, WHEN GÜNTER Wächtershäuser first put forth the idea in academic journals that life had originated in the deep ocean, nobody paid attention. After all, Wächtershäuser wasn't an academic, he wasn't a professional scientist. He was a lawyer practicing international patent law in Munich, Germany. And there was no getting around the fact that his argument sounded nuts. Wächtershäuser thought that all life on Earth started from a chemical reaction between two minerals, iron and sulfur. This reaction touched off a metabolic process that created a single molecule. Once that process was under way, it fueled the creation of more complex molecular compounds, which would evolve into life forms, and, eventually, us.

You, me, the birds and the bees, the bushes and the trees — we all came from rocks. And these rocks came from the black and boiling water of hydrothermal vents. Wächtershäuser called it the iron-sulfur world theory.

To understand just how controversial Wächtershäuser's theory is, keep in mind the generally accepted view of the origin of life at the time. In the 1980s, most scientists subscribed to some form of the soup theory. This theory — explained here in the most basic terms — argued that around four billion years ago, chemicals in the primordial sea, the "soup," with input from energy sources like lightning, reacted to form the first organic compounds. These compounds eventually formed more complicated structures, which eventually grew into early kinds of life.

Wächtershäuser, who received a doctorate in organic chemistry, believed in the soup theory in his academic career. Then he got fed up with academia and pursued chemistry as a hobby while practicing law. During that time, he took the soup theory apart and discovered numerous holes.

For instance, the soup theory posited that chemicals mixed freely in the water and air to make more complex molecules. The problem was, as Wächtershäuser thought, chemicals don't stay together for long in a free-floating, three-dimensional environment: On the surface of rocks, however, chemicals were stable and could combine and grow into more complex forms.

In most soup-theory models (and there are many), cell membranes are believed to be the first element of life. But if that was the case, how did food get through the cell membrane and into the cell? Without fuel, the cell had no way of staying alive. Strike two against the soup.

None of these issues affected iron sulfide. In the hot, pressurized waters of hydrothermal vents, chemicals could combine and recombine on the two-dimensional surfaces of these minerals relatively quickly and easily.

Wächtershäuser argued his case in obscurity for years. In 1997, he and a researcher at the Munich Technical University decided to test the hypothesis by combining the gases found in deep-sea vents with iron and nickel sulfides. The result surprised everyone. From this simple mixture, an active form of acetic acid, an organic compound made of two bound carbon atoms, was produced. This form of acetic acid can react with other chemicals, meaning that the reaction may have been the first step in the origin of life. The results of this experiment were published in the April 1997 issue of *Science*.

In April 2000, researchers at the Carnegie Institution of Washington's Geophysical Laboratory took the iron-sulfur theory one step further. They not only combined the hydrothermal gases and iron minerals that Wächtershäuser used in 1997 but put everything in a steel pressure chamber that mimicked the pressures of water in the deep ocean.

"We came across an unanticipated result," George Cody, the lead researcher on the study, told the *New York Times*. The pressurized mixture produced pyruvate, a molecule made up of three linked carbon atoms. Pyruvate is a key component of living cells and a building block for multiple organic compounds.

Wächtershäuser claimed victory, writing that the experiments "greatly strengthen the hope that it may one day be possible to understand and reconstruct the beginnings of life on earth."

Years later, further tests of the iron-sulfur theory produced more startling revelations. In an article published in the January 2003 issue of *Philosophical Transactions of the Royal Society*, researchers Michael Russell and William Martin argued that certain structures of hydrothermal vents made perfect incubators for organic molecules. Russell demonstrated his point as early as 1997 by dissolving hydrothermal gases in the lab and then adding an iron-rich solution to the mix. Within a minute, an inch-high honeycomb of compartments emerged. Even more amazing is that the membranes of the newly formed rocks separated two solutions with different ion concentrations, creating a voltage across the membrane of about six hundred millivolts. This voltage, which lasted for several hours, was about the same as the voltage across cell membranes and could be enough to support the formation of compounds.

"It's a little bit of rock that reminds us where we came from," said Russell.

If it is true, the iron-sulfur world theory suggests that life not only could have started in hydrothermal vents but that it had to have started there. No other environment had the pressure and chemical components needed to produce organic compounds that led to early life. The process in the vents was so reliable and consistent that life most likely emerged from hundreds or thousands of vents at around the same time — trillions of different cells replicating in the boiling water of the Earth's core across the seafloor.

A people born from the entirety of the world's oceans.

IT'S THURSDAY, TWO DAYS AFTER we've lost sight of the landers, and I'm standing on the aft deck of the R/V *Falkor*. A dozen crew members in yellow hardhats, cargo shorts, and knee-high rain boots are tugging at a half-dozen ropes that spool out from deck into the roiling waters. The ropes are wrapped around a white rectangular box about the size of a small refrigerator, which is floating about a hundred feet behind deck. The box's name is LEGGO, our expedition's last surviving deep-sea lander, and it's just resurfaced after a sixteen-hour dive. Bartlett's colleague, David Price, sent LEGGO down to Challenger Deep with a net filled with chicken in the hopes of attracting some hadal life forms to bring back to the surface. It worked. Sloshing around in the net is a mound of pinkish-white life forms, what could be a new species of hadal life. The crew needs to act fast. The longer it takes them to pull LEGGO back to deck, the more likely they'll lose the precious cargo to the rocking waves.

"Careful! Careful now!" yells a man with a British accent. Attached to one of the ropes is a Portuguese man o' war, a vicious jellyfish-like animal which can inflict an extremely powerful sting that's sometimes lethal. "Don't touch it!" someone yells. "Get it off the rope!"

The process is agonizingly slow, despite the rush. If LEGGO gets too close to the R/V *Falkor* before it's hoisted to deck, a swell could smash it to the side of the hull, losing all the hadal life forms and smashing the camera equipment.

A crew member lowers a crane over the stern and pulls up the slack of the ropes. Now LEGGO hovers a few feet above the surface. But the danger hasn't passed. New swells lift up the lander (which weighs several hundred pounds) and drop it, until it's swinging violently from side to side like a wrecking ball. With each swing, a few clumps of the pinkish-white life forms flop out of the net and back into the ocean, disappearing on their way back to their hadal home.

Eventually, the crew manages to guide LEGGO back to deck and secure it with cables. A spontaneous round of applause breaks out. "Okay, scientists," says the British man. "Come and get your stuff." The team jostles over to the net, opens it, then carefully dumps its contents into test tubes, plastic dishes, and bags. The life forms — whatever they are — look like shrimp: each is about two inches long, pink, and even from twenty feet away, I can smell their foul, sulfurous odor. The science team rushes the samples to a wet lab and begins the hour-long process of labeling the animals and storing them so that they won't deteriorate before the team can get samples back to the laboratory in San Diego and determine exactly what they just discovered: a new species, a new life form, perhaps some new disease that will destroy all mankind? "We should know more in a few weeks," says Bartlett, grinning, as he rushes around the lab. "You know, if we can discover something new here, something nobody else knows, that will be very exciting."

BEFORE WE HEAD BACK TO port, there's one other thing left for me to do.

When I started this project in 2012, my goal was to participate in as much deep-sea research as I could. I was writing about the human connection to the ocean, after all; to not see and feel that connection myself seemed dishonest and wrong. This wasn't possible with research at all the depths. For instance, I couldn't exactly see or feel the deep water at ten thousand feet, but I could at least see those mysterious animals that dwelled down there when they came to the more accessible waters of the surface. And I did see them, and was lucky enough to dive with them and feel them see me with their bone-rattling echolocation.

The hadal zone was an exception. No hadal animals can swim to the surface; many don't even make it up to twenty thousand feet. And yet, I was still determined to personally experience these depths in some way. At minimum, I wanted to leave behind a memento.

So, while Bartlett and his team are busy processing their hadal catch, I walk back to my room, grab a ziplock bag from my suitcase, then climb the gangway to the roof of the navigation deck. Inside the ziplock bag is a tennis-ball-size white plastic container of Daggett & Ramsdell knee and elbow skin-lightening cream, extra-strength formula. Inside this container is a keepsake I'll be dropping down to Challenger Deep.

I've never used elbow whitener and didn't even know such a thing existed until last year when I built my first deep-sea memento, which I dropped into the Puerto Rico Trench, the Atlantic Ocean's deepest waters. It turns out that the round 1.5-ounce jar that holds elbow whitener is ideal for the outside of a double-hulled vessel capable of surviving the crushing pressures of the hadal zone. And the tiny glass container that holds Golden Cup Balm, a Thai muscle-pain-relief ointment, is an excellent pressure chamber.

I emptied the contents of both products, placed my memento inside the balm container, dropped this into the larger elbow-whitener jar, filled both with silicone oil, and sealed them shut. A perfect fit. Even the smallest air bubble could implode this vessel on its downward journey. The silicone oil removed any air and also protected the whole apparatus from the crushing pressure of 35,000 feet. Any liquid works; I used silicone oil because it wouldn't harm the delicate electronics I've placed inside.

Now, on deck, I remove the white plastic container from the ziplock bag, arch my right arm back, and throw it over the side of the *Falkor*. I watch as it crashes down in a wreath of foam, bobs for a second, then slips beneath the ocean's surface, drifting down in slow motion, inch by inch, foot by foot, until there's nothing left but a twinkling of white against a blue and empty space. But now, unlike a year and a half ago, I know there is nothing empty about this space that surrounds me.

THERE ARE MORE LIVING THINGS here and more different kinds of life in the ocean than anywhere else in the known universe. And as my slapped-together memento hovers above the deep ocean floor like a satellite miles high in the sky, it strikes me, and not for the first time, that the farther we descend into the light-less depths of the sea, the closer we get to understanding our origins — our amphibious reflexes, our forgotten senses, where we came from.

Inside the white plastic container I just dropped is a digital copy of the book you've just read. These words you're reading are drifting into the planet's deepest waters, sinking miles from the sunlit surface. But they're not lost to a distant, alien world. The sea is where all life began billions of years ago, and where all living things will eventually return.

Hours later, as we begin the long haul back to port, I imagine the container touching down, silently, on the sunless valleys and hills of the planet's deepest seafloor, where it will stay for the next few thousand years, getting softly dusted by the never-ending snowstorm of microscopic skeletons that will one day cover some future Earth.

As quickly as it began, the downward journey has ended. We've finally made it home.

# Ascents

## -35,850

The LEGGO lander had made it down just over 35,850 feet, what is most likely the deepest drop ever completed. And the "shrimp" LEGGO pulled up in its nets are most likely the deepest life forms yet captured. Bartlett and his team are now dissecting the samples and extracting microbes from the creatures' guts. In the next few weeks, they'll put the microbes in canisters, pressurize them to the hadal depths, and try to regrow them into cultures. "We want to see how these things can survive in such incredible pressures," said Bartlett when I talked to him on the phone. "We just don't know about the potential for life down there – or anywhere."

David Barclay's deep-sea recordings were equally revelatory. With his remaining lander, Deep Sound 2, Barclay gathered sound recordings from more than five and a half miles below the surface. He discovered that sound generated at the surface couldn't penetrate below 16,000 feet; instead, it bounced back upwards. As a result, the ocean at these depths was extremely quiet. Although researchers had theorized that sound would behave this way, nobody knew for sure. Nobody had recorded there before. Barclay is currently processing the recordings below 16,000 feet to see if oceanic animals might be using this quiet zone to communicate, or more.

And although Barclay never recovered Deep Sound 3, all was not lost. While analyzing the recordings from hydrophones tied to the side of R/V *Falkor,* Barclay noticed enormous noise spikes 247 minutes after Deep Sound 3's launch. The first spike was the sound of Deep Sound 3 imploding at 27,000 feet; the second was its echo off the seafloor at more than 35,000 feet — the deepest implosion ever recorded.

## −10,000

Fabrice Schnöller's "big and important" recording that he made off the side of the motorboat in Trincomalee may actually turn out to be big and important. Schnöller believes he captured a "gunshot" — the rarest and most confounding of sperm whale vocalizations.

The sound is connected, scientists think, to a sperm whale's hunting technique. Unlike a baleen whale, which uses a hairlike substance to filter plankton from the water, a sperm whale has about forty teeth along its lower jaw. Whalers assumed sperm whales used these teeth to attack prey, but research suggests otherwise. Examinations of dead sperm whales' stomachs (a sperm whale has four) show that they don't chew their food. Giant squids, a primary food source for them, swim up to thirty-five miles per hour and grow more than sixty feet long. The sperm whale has a top speed of about twenty-five miles per hour. How could it catch, let alone kill, a giant squid without biting it as it passed by? And what good are teeth when you don't chew your food?

Some researchers, including Schnöller, believe sperm whales use their teeth as little antennas to assist in echolocation, or even holographic communication. To hunt, sperm whales probably use super-powerful gunshot clicks to stun or kill prey before they consume them.

There have been only two recordings of sperm whale gunshots ever made: in 1987 and 1999, both in Sri Lanka, both by scientists. Schnöller believes the recording he made with his pasta strainer, broomstick, and homemade hydrophone rig is the third. "I know it's a gunshot," he wrote. "I'm now going to do proper scientific protocol to prove it."

## −2,500

Karl Stanley has had his share of disputes with authorities in the past, and many travel agencies continue to refuse to promote his tours. However, in the last few years, he reports that most of these problems have subsided and his presence on Roatan is welcomed by locals and authorities. He told me after our trip that in 2011 he took the president and vice president of Honduras down for a ride to −660 feet. In March 2012, he completed his longest dive ever, spending twenty-four hours submerged along the Cayman Trench. Fourteen of those hours were spent below 2,400 feet.

As of August 2013, the Roatan Institute of Deepsea Exploration and the general public's deepest diving submarine were still in operation.

## −1,000

In September 2013, after spending five years and thousands of dollars building his own dolphin and whale recording devices, Fabrice Schnöller launched a new company called Click Research (www.click-research.net), which provides citizen-scientists affordable gear to do DIY oceanic research. Click Research's products, which were developed with DareWin engineer Markus

Fix, include a shark-monitoring device, a transmitter that sends live whale songs to your home stereo, and a dolphin vocalization analyzer that identifies dolphins by their signature whistles. Schnöller hopes to one day use a souped-up version of the dolphin analyzer to translate dolphin whistles into English. "In a couple years, possibly," he told me. "But it is very complicated, you know."

Schnöller and the rest of the DareWin crew plan to take this equipment – and the thirty-nine-speaker holographic-communication rig – on a two-week expedition to a deserted beach off the coast of the Arabian state of Oman in 2016. It will be the first field test of holographic communication ever attempted.

"You must come!" he yelled over the scratchy phone line. "It's going to be insane!"

## −800

A year after his failed world-record no-limits dive to eight hundred feet in Santorini, Greece, Austrian Herbert Nitsch is still suffering from neurological problems. Nitsch can't remember names well; he has trouble talking and walking. His voice is shaky, and he has lost much of the mobility in his right arm. He continues to improve every day, he says, and is now working on his rehabilitation training with the same enthusiasm and determination he used to approach no-limits diving. He's even begun freediving again, but only to a depth of about ten feet.

Nitsch now focuses much of his time on ocean conservation. He joined world champion surfer Kelly Slater and legendary freediver Enzo Maiorca as a member of Sea Shepherd's Ocean Advocacy Advisory Board to help end the slaughter of wildlife and the destruction of the world's seas.

When a reporter asked Nitsch what he considered to be the current world record in no-limits diving, Nitsch responded, "To tell the truth, I don't care now."

In the wake of the success of the SharkFriendly pilot system in December 2011, local authorities began a tagging program of their own, and over the course of the next year they tagged acoustic transmitters on more than a hundred bull sharks along Réunion's west coast. While the tagging system successfully tracked the movements of bull sharks, it did little to stop them from attacking swimmers. From January 2012 through August 2013, Réunion Island has suffered three more fatal shark attacks. The authorities closed off Réunion's beaches and, at last report, were planning to kill off ninety bull sharks.

Fred Buyle predicts it will do little good and will most likely provoke even more attacks in the future. "Surfers know these are dangerous situations, but they go out anyway, then they blame the sharks," he said. "People need to learn that when they are in the ocean they are swimming in nature. The only solution here is education: Don't swim in cloudy water. Don't swim after a big rain. Don't swim near a river. But nobody listens."

COINCIDENTALLY, THE SOLUTION FOR RÉUNION'S shark problem may come not from Buyle and Schnöller but from Jean-Marie Ghislain, the shark researcher I first met in Cape Town through Hanli Prinsloo. In September 2013, Ghislain visited me in San Francisco for three days and explained his proposal. He had been consulting with a Belgian company, AquaTek, to create a shark-deterrent system called Shark Repelling System (SRS). SRS would disrupt sharks' electroreceptive sense by blasting the water with a magnetic field. In dozens of tests using both captive and wild sharks, it had a 100 percent success rate, often scaring sharks off from hundreds of feet away. The sharks, Ghislain said, suffer no ill effects, and because the system affects only animals with electroreception—sharks and rays—other sea life can pass through the SRS unharmed.

AquaTek is in negotiations with Réunion authorities to install the SRS at Boucan Canot, Saint-Gilles, and other beaches along Réunion's west coast in 2014. "This could be a big paradigm shift in our approach to sharks," said Ghislain. "It has a very good chance of saving them . . . and saving us."

## −300

Dave King, the British diver who went into cardiac arrest after he completed a monofin attempt to 335 feet at the world free-diving championship in Kalamata, Greece, fared much better. He has suffered no ill effects from his blackout. "I am not a reckless diver," he wrote a few months after the accident. He claims the blackout in Greece was his only one in ten years of freediving. He argued that his work schedule doesn't allow him to train as much as other elite divers and that he had time for only three dives before the competition. "I got to 102 meters, equalizing easily," he wrote. "I just had problems as I reached the surface."

## −60

In June 2011, a month after I visited the Aquarius Reef Base in Key Largo, the National Oceanic and Atmospheric Administration (NOAA), the federal agency that manages Aquarius, cut funding for the base and canceled all upcoming missions. Soon after, the world's last underwater habitat was shuttered.

Then, in early 2013, Florida International University negotiated with NOAA to take over operations. In September 2013, Aquarius was reopened and aquanauts were back inside the habitat, sitting around half robed in the cold, wet kitchen, buzzed off nitrogen, eating flattened Oreos, and uncovering the secrets of the ocean and ourselves sixty feet below the surface.

# Epilogue

FIVE DAYS BEFORE MY DEADLINE to deliver the final draft of this book, on November 17, 2013, I get an e-mail from Fred Buyle. "Remember the conversation we had, James, when I told you that someday we'll see someone die in a freediving competition," he wrote. "Today, it happened."

At 1:45 p.m. local time, Nicholas Mevoli, a thirty-two-year-old from Brooklyn, died of complications from lung damage shortly after completing a 236-foot no-fins dive. The dive happened during Vertical Blue, an annual freediving competition hosted by William Trubridge at Dean's Blue Hole in the Bahamas.

Mevoli was new to the sport. Just eighteen months earlier, in May 2012, he had made his competitive freediving debut with a monofin dive to three hundred feet—an unheard-of depth for a beginner. The following year he competed in dozens of competitions, attempting even greater depths. He often blacked out. He frequently resurfaced bleeding from his nose and mouth. He tore his lungs repeatedly, which caused him to spit up blood days after competitions. Mevoli brushed these warning signs aside. He kept diving. And he started breaking records.

On November 16 at Vertical Blue, Mevoli attempted a 314-foot dive, a U.S. national record in the discipline of free immersion. He made it down to 260 feet, then suddenly turned back. Safety divers had to pull his unconscious body to the surface. When Mevoli came to, blood dripped from his mouth. He thrashed in the water angrily and cursed at himself. "Numbers infected my head like a virus and the need to achieve became an obsession . . . Obsession can kill," he had written in a blog months earlier. Unfortunately, Mevoli didn't heed his own advice.

The next day, November 17, still reeling from the blackout, Mevoli announced that he would break another U.S. record, this time in the no-fins discipline, the most demanding in freediving. He was trying for a depth of 236 feet. At 12:30 p.m., he slipped on his goggles, took one last breath, and kicked his bare feet along the guide rope. Moments later, Mevoli disappeared in the shadows of the deep water.

A judge on deck followed his descent. Mevoli dove quickly, passing 50, 100, 150, 200 feet. Then, at 223 feet, just a dozen feet short of his target depth, he unexpectedly stopped. Moments passed, and Mevoli didn't move. Some time later, he started his ascent, stopped again, turned around, and forced himself back down to the plate at the end of the rope. Fifteen divers, judges, and medics waiting on deck at the surface winced. They knew this was a reckless decision. Mevoli grabbed a ticket off the plate at 236 feet and scrambled back up the rope.

Somehow he was still conscious when he reached the surface. He flashed an okay sign and tried to complete surface protocol by saying, "I'm okay." But the words never came. Seconds later, he blacked out. The medics lugged Mevoli's unconscious body to the dive platform and began emergency resuscitation procedures. Blood poured from Mevoli's mouth. His pulse fluttered. Fifteen minutes later, the pulse disappeared, and the medics cut off his wetsuit and began vigorously pumping his chest. They injected him with adrenaline. The resuscitation attempts went on for almost ninety minutes. They put him in a station wagon and drove him to a nearby hospital, where the paramedics continued to work on him and a local doctor pulled about a liter of fluid from his lungs. Soon after, Mevoli was pronounced dead.

IT WAS THE FIRST DEATH in twenty-one years of AIDA-sanctioned freediving competitions. Buyle was both sad and furious.

After the death, Buyle posted an open letter to Nektos.net, his website, describing how the new generation of competitors had become disconnected from the sea, themselves, and the true nature of freediving.

Buyle mentioned that it took him ten years of conditioning and constant diving to make it to the depths Mevoli was diving to after just a year and a half. The new divers were going far too deep, too fast, and they were skipping what Buyle called "the adaptation phase needed to survive a deep dive." They were also putting themselves at grave risk. "At some point I started to be worried that a serious accident could happen," Buyle wrote. These competitors, he said, were "looking deliberately for trouble."

NICHOLAS MEVOLI'S DEATH MADE international headlines. Three days after I received Buyle's e-mail, I was asked to comment on the tragedy for Al Jazeera television. The next day I appeared on National Public Radio's *Weekend Edition*. In the two years since I'd been introduced to freediving, I'd somehow become an authority on it. This was flattering, but it also seemed absurd to me. I'd attended all of two freediving events, but that was two more than most journalists. And I'd certainly seen enough to form an opinion. Above all, I wanted to set the record straight.

What I told NPR and Al Jazeera and the rest of the press, parents, friends, and recreational freedivers who called and wrote me that week — and what I've tried to very clearly explain in this book — was that competitive freediving was profoundly different from the kind of freediving I'd researched and trained to do.

Most competitive divers are blind, numb, and dumb to the ocean environment. They go against their bodies' instincts, ignore their limits, and exploit their amphibious reflexes to the breaking point. They do this to dive deeper than the next guy. Sometimes they make their target depth; sometimes they don't. Sometimes they resurface unconscious or paralyzed or worse.

The freediving I learned from Prinsloo, Buyle, Schnöller, Gazzo, and the ama was the opposite of this egocentric, numbers-driven approach. To this group, freediving was about connecting with the underwater environment, looking more keenly at your surroundings, focusing on your feelings and instincts, respecting your limits, and letting the ocean envelop you — never forcing yourself anywhere for any reason. This was a spiritual practice, a way of using the human body as a vessel to explore the wonders in the Earth's inner space.

Freediving was also a tool. It gave my teachers access to the ocean that nobody else had. Using it, this group was helping to disprove many long-held assumptions about the ocean and its inhabitants. (Sperm whales don't want to eat us; dolphins are trying to talk to us; sharks can become docile and playful when approached on their terms.) This group was also contributing to larger discoveries that could one day have a significant impact on the way we view life on Earth, and our place within it.

WHEN AND WHAT WILL WE know for sure? If history is any guide, Schnöller's theory of click and holographic communication will take years, perhaps decades, to be proved or disproved. The big, paradigm-shifting scientific discoveries always do.

It wasn't until the 1980s, some twenty years after Friedrich Merkel's experiments with European robins, that scientists proved the existence of magnetoreception. Patent lawyer Günter Wächtershäuser toiled in obscurity for a decade until his iron-sulfur world theory was tested and gained scientific support and respect. And so on.

I realize that the idea of talking to a dolphin or exchanging three-dimensional images with sperm whales sounds mad. It certainly sounded crazy to me when I started this project, and I still want to pull out my notes whenever I'm chatting with a stranger about it, just so I'm fully armed with the facts.

But the reality is, we don't have time to doubt Schnöller or the others plying the deep. The ocean is changing. The seas are rising. Coral is dying off and will probably disappear altogether in fifty years. Environmental hazards abound on the open sea — oil spills, trash, sound pollution, nuclear waste, and so on — and all or some of it is killing whales and dolphins and species we don't even know about. One hundred million sharks are killed in the world's oceans every year. These animals may be gone before we even have a chance to fully understand them.

And whatever we learn about them will lead us, undoubtedly, back to ourselves.

Over the past two years, it's become clear to me that we don't know what we are yet. And now the truth is like a bell constantly ringing in my ears.

I heard it for the first time, very clearly, in Sri Lanka.

IT WAS THE LAST DAY we were together on the beat-up motorboat; Schnöller and Prinsloo had been bickering since dawn, the temperature had spiked to 110 degrees, and signs of genteel mutiny were everywhere. We were floating in the deep water of Trincomalee Canyon and had given ourselves until noon to see whales. Noon came and went. No whales. No sign of land either. Nothing but dead-calm water in all directions, and sun.

I suggested a swim. Cameras, itinerary, strategies, plans, and talk will be left on deck, I added. We'd just dive together, just this once, for no other reason than that it sounded like fun. Everyone agreed. We slipped on our gear and jumped in, one by one.

Within moments they were gone, through the doorway to the deep, and beyond.

I took a breath, grabbed my nose, upturned, and dove down to join them. I saw Guy Gazzo first. He was hovering in neutral buoyancy, his fingers interlocked behind his head as if he were napping on an imaginary chaise longue. Schnöller stretched out fully beside him, spinning in lazy horizontal circles like a thrown baton. Below them, some seven stories from the surface, Prinsloo and her beau, Marshall, swam double helixes around each other until they all but faded away into the shadows.

*What are we?* I thought to myself. And with every breath I hold, I still wonder.

# Acknowledgments

TWO YEARS AGO, I WAS dangling my feet off the prow of a sailboat, manically scribbling in a notebook, attempting to describe the tension, elation, horror, and profundity of competitive freediving. At the end of the first day of competition, all I had were a few names, times, and quotes. Just the facts. Freediving left me utterly speechless, and subsequently noteless.

That night, Alex Heard, my editor at *Outside* magazine, called to check in. I remember mumbling something to the effect of "freediving is like being in space, but it's in water; like flying but you're diving. It's the most . . . the worst . . . the best . . . the bloodiest." Alex must have hung up more confused than he was before he called. But over the next several weeks he helped me get the words out. The resulting article, published in the March 2012 Adventure Issue, sparked this book. I'm very grateful for Alex and the crew at *Outside* for sending me overseas for ten days to cover a sporting event I knew nothing about.

Field research is hard. Field research at sea is harder. Field research miles off the coast of a developing country in a rickety boat with a bunch of do-it-yourself researchers on shoestring budgets using slapped-together equipment to study the ocean's largest predators often verges on suicidal. That nobody was seriously injured during the writing of this book is a testament to the ace improvisational skills of the crews with whom I was lucky enough to share the past year and a half. Or maybe it was just dumb luck.

Thank you, Fabrice Schnöller, Hanli Prinsloo, and Fred Buyle, for letting me into your watery world. Thank you for saying things like "Dolphins are usually friendly, but sometimes they might try to rape you" and then yelling at me to get in the water . . . with dolphins. Thank you for lying when you said there weren't any sharks at the beach we had just dived at. Thank you for taking my arm and pulling me down to have my bones rattled by toothed whales. Thank you for not mocking my French more than three times a day. Without your persistent prodding, I doubt I'd ever have gotten wet.

Although I'd lived around the ocean my whole life, I was pretty clueless about what happened below the surface. Dozens of kind, patient, and brilliant marine scientists helped light the way to the darker depths. They responded to my e-mails, called me back, spent hours explaining stuff that took me months to fully understand. And they did it for nothing more than a lame verbal remuneration that usually included phrases like "Seriously helpful"; "Whoa! Excellent"; "This is really great." I'm talking about Stan Kuczaj at the University of Southern Mississippi, Saul Rosser at Advanced Diving Systems, Alan Jamieson at the University of Aberdeen, Fabienne Delfour at the University of Paris, Robert Vrijenhoek at Monterey Bay Aquarium Research Institute, Bart Shepherd at the California Academy of Sciences, John Bevan at Submex, and Douglas Bartlett and Paul Ponganis at the Scripps Institution of Oceanography. And lest I forget the brainy and bad-ass Kim McCoy at Ocean Sensors, who shared his fat Rolodex and decades of oceanic expertise more than a few dozen times. (Kim is a cheap date: Just buy him an espresso at a café in downtown La Jolla and he'll tell you anything you want to know, all in double-time.)

It's a fine line between editing and rewriting. Luckily for me, Danielle Svetcov at Levine Greenberg Literary Agency crossed that line many times when I needed it most. A literary agent who picks up the phone is a gift. One who knows how to edit is a treasure. But what to call one who stays up until three a.m. rereading dozens of pages of draft chapters that are due to an editor the next morning, and does this not once, not twice, but a zillion times? And does this with an unshakably blithe spirit? And then *still* returns your phone calls the next day? I call this person ridiculous. Thank you, Danielle Svetcov. Now please get some sleep.

If you took all the ink in all the acknowledgments of all the books published in the past two decades that praised the lethal wit, editorial prowess, and philosophical erudition of Eamon Dolan and then spread that ink out on a single piece of paper and laid that piece of paper over land, you'd cover the surface of Guyana. (Look at a map and you'll discover that Guyana is quite large, some eighty-five thousand square miles.) The rumors are true: Eamon Dolan is the real deal. His persistent and patient support and expert advice was unwavering from the get-go, even through the dark hours of what became an arduous sprint to the finish line. So there it is, more ink. So predictable, I know, but then again, so well deserved. (Thank you, Eamon: I promise never to use the words *incredible* and *tremendous* to describe anything ever again.)

Will blood squirt from unprotected human eyes at −5,000 feet? How long does it take for a duck to drown? What happens if you pee in an atmospheric diving suit? This is the kind of unpleasantness that awaited Julie Coombes's in-box day after day for the better part of a year. Julie assisted in the historical and scientific research of this book and did so with unblinking assiduity (a phrase she would rightfully have edited to *really well* had she gotten her mitts on this page). She saved me from factual missteps more than too many times. Thank you, Julie Coombes, for your good humor and better hourly rates.

There're dozens of other people who assisted in less direct ways in the research and writing of *Deep*. Some of them became minor characters; others provided invaluable information; a few just bought me an occasional beer and listened to me complain about the sucky movie selection on United's international routes. They are Markus "Dream Killer" Fix, Max Landes, Stig Severinsen at Breatheology, Bertrand Denis, Captain Jose, Steven Keating at MIT Media Lab, David Lang at OpenROV, Marc Deppe at Triton Submarines, Tad Panther, and Adam Fisher. Daniel Crewe at Profile Books (UK) provided unabashed encouragement early on and ace edits at the end. Thank you, Daniel.

I'm also very appreciative of the support and Swiss-neutral professionalism of Emmanuel Vaughan-Lee and the crew at Go Project films. These guys got the short end of the Sri Lankan sperm whale stick, but they never squealed. Add to this list Jean-Marie Ghislain, whose underwater photographs are some of the best I've ever seen and whose keen skills in the Belgian art of diplomacy kept our sinking trip afloat.

The stunning and otherworldly photographs that appear in the middle pages of this book come courtesy of Fred Buyle (nektos. net), Jean-Marie Ghislain (ghislainjm.com), Yann Oulia, Olivier Borde (olivierborde.tumblr.com), and Annelie Pompe (annelie-pompe.com). Merci les gens merveilleux qui sont français, belge, et ceux qui ne sont pas français!

While you're looking all that up online, please gambol over to Hanli Prinsloo's I Am Water (iamwater.co.za) and DareWin (darewin.org). Both of these organizations are using a hands-on, direct-action approach to ocean exploration and conservation, and, so far, it's working. And they're doing it on the shortest of shoestring budgets. Contact them to get involved.

Will Cockrell at *Men's Journal* somehow convinced his boss to send a writer he'd never worked with to an island neither of us had heard of to cover a project that had a very good chance of failing. That project, SharkFriendly, was the focus of the feature "The Shark Whisperer," published in the June 2012 issue of *Men's Journal*. If it weren't for Will, I would never have made it to Réunion and never have met Schnöller; the past two years of deep-water delving would very likely never have happened. Thank you, Will and the *MJ* staff, for rolling the dice. (And to the rest of you: Don't let the shiny gear and buff bros that sometimes plaster the covers fool you; *Men's Journal* is one of the best magazines going.)

If you haven't figured it out already, freediving can be a dangerous hobby and a lethal sport. Many freedivers delude themselves into thinking otherwise. As a result, many die every year in accidents that are easily preventable. The crackerjack, no-bullshit approach offered by Eric Pinon at Performance Freediving International and Ted Harty at Immersion Freediving has been a lifesaver for me and thousands of other beginning divers. If you want to go deep, start at the top. See these guys. And remember: Know your limits. Never dive alone. Always stay in control.

William Trubridge will certainly not approve of this book, and I certainly do not approve of William Trubridge's approach to freediving, but I'd still like to thank him for the five hours he spent talking with me in Greece and giving me the inspiration to take the plunge.

That creep you've seen splayed out beneath a row of chairs with a T-shirt over his face, sleeping through a seventeen-hour layover in the Dubai airport? I was that guy for a year and a half. Needless to say, I wasn't home much. Brent Johnson and Maile Sievert watched my dog, Face, during the months away. Amanda at Amanda Bilecki Moler Acupuncture fixed my broken body when I returned. Circle Community Acupuncture kept things running (thank you, Jenn, David, and Melissa).

My mother warned me seven years ago to never quit my day job. She was wrong about that, but she has been right about so much else. Thanks, Mom. I promise to keep forwarding my Kayak itinerary confirmations for any upcoming trips.

*Deep* was written between stints at the San Francisco Writers' Grotto, various rental cabins in Inverness, California, and the second-floor desk between the stacks of decorative arts books at the Mechanics' Institute Library in San Francisco.

*Ĉi tiu libro estas dediĉita al tiuj,*
*kiuj klaki la Majstro Switch.*

This book is dedicated to those who flip the Master Switch.

# Notes

−60

PAGE

18 *Coral has no eyes, no ears, and no brain:* Since corals' annual mass spawning was first discovered, in 1981, it has stumped scientists. Coral is a primitive animal with no sense of sight or hearing, and yet it manages to communicate with other corals in ways more sophisticated than our own.

In 2007, a group of Australian and Israeli researchers tried to find out how. They discovered that coral had a gene called *CRY2* that allowed it to distinguish between subtle changes in light. Many plants and animals, including humans, share the *CRY2* gene, which is associated with monitoring levels of light as well as sensing subtle shifts in magnetic fields. In humans, *CRY2* proteins help set circadian rhythms of sleep and may also be related to depression and mood disorders. Corals were using the gene as the tiniest and most primitive of eyes.

"This particular gene allows the coral to sense blue light and to actually work out what phase the moon is in," explained Bill Leggat, a coauthor of the study, in the October 22, 2007, issue of *Science.* And using these *CRY2* genes, Leggat and the scientists believed, corals might be able to sense the passing of seasons and so time their mass spawning to a particular light on a particular day. The corals weren't telepathic at all; they were just taking cues from the sky and counting the days.

While it came as an epiphany to some, Leggat's theory pushed against many reports from the field.

For instance, the Leggat report didn't consider that corals spawned in synchronicity even when there was no light. In other words, a chunk of coral species totally hidden from natural light would still spawn at the same time as other corals a hundred feet deep and hundreds of miles away. Aquarium owners around the world often witness this phenomenon.

Theories aside, in 2007, the mere fact that corals had the *CRY2* gene was newsworthy. For Leggat and other scientists, it was an example of how closely humans were tied to the ocean and to even the most primitive animals within it.

"*CRY2* were preserved for hundreds of millions of years before being inherited by corals when they developed about 240 million years ago, and are still found today in modern animals and humans," said Leggat. Ove Hoegh-Guldberg, director of marine science at the University of Queensland, said of the *CRY2* discovery, "They are an indicator that corals and humans are in fact distant relatives, sharing a common ancestor way back."

The human connection to the ocean, it appears, extends even to the crusty clumps of white rock on the seafloor.

−300

28   *life-lengthening effect on humans:*  The first experiments into animals' amphibious reflexes were conducted twenty years earlier by French physiologist Paul Bert. In the 1870s, Bert started drowning ducks and chickens and timing how long it took each to die. The ducks consistently lasted from seven to sixteen minutes underwater, while the chickens survived for only three and a half. From a scientific standpoint, this made little sense. The animals were biologically very similar — with the same lung volumes, weight, and circulatory systems — and yet the water seemed to extend the life of ducks and drown the chickens more quickly.

Bert continued looking for answers in a series of bizarre experiments he called "death in closed vessels," a horrible if illuminating exercise. He bled ducks to the point where they had the same blood volume as the chickens and then drowned both to see which would die faster. (The chickens still died two to three times faster than the ducks.) He crammed newborn kittens in bell jars, sealed the jars, then timed how long it took them to die. (They died in about the same amount of time as strangled grown cats.) He drew blood from a dog, killed it, ran an electrical wire through its mouth and out its anus, electrified the corpse, and checked to see if the oxygen levels changed. (They didn't.) He urinated in various bottles and exposed the bottles to different air pressures for up to several days. The results were, in Bert's words, "completely turbid, very alkaline, horribly foul."

Six hundred and fifty experiments later, Bert had killed off dozens of dogs, sparrows, rats, cats, rabbits, kittens, owls, chickens, and ducks and had saved himself a few trips to the latrine, but he was no closer to understanding why ducks were able to survive underwater longer than chickens or other animals. What he did discover, however, was that breathing high concentrations of oxygen could lead to oxygen poisoning (later called the Paul Bert effect). Bert's 1,050-page book *Barometric Pressure: Researches in Experimental Physiology* became an instant classic when it was published in 1878 and paved the way for scuba diving and high-altitude flying in the next century. Today, Bert is considered the father of aviation medicine.

29 *He'd seen the same thing happen in deep-diving seals decades earlier:* By the late 1960s, research in the physiology of diving animals became increasingly bizarre — and grotesque. None topped those by Robert Elsner, a marine animal physiologist. Elsner's battery of experiments, published in 1969 by the *Yale Journal of Biology and Medicine,* included cutting open the bellies of pregnant sheep to check for maternal and fetal responses to asphyxia. Elsner, along with colleagues D. D. Hammond and H. R. Parker, also traveled to Antarctica and performed similar experiments on Weddell seals. What they found was that sheep and seal fetuses both responded to asphyxia in similar ways: the heart rates dropped and blood was shunted to the vital organs.

44 *nitrogen narcosis:* Nitrogen gas can build up in the blood to dangerous levels during very deep freedives assisted by heavy weights or machines or when a diver descends below a hundred feet several times in quick succession over a period of hours. Ancient South Pacific pearl divers who dove forty to sixty dives a day, sometimes

to as deep as 140 feet, suffered a severe illness they called taravana, the symptoms of which—dizziness, numbness, visual confusion—closely resembled what later became known as decompression sickness. In the 1970s, Dr. Edward Lanphier showed that decompression sickness could easily be avoided by either diving to shallower depths or spending twice the amount of time at the surface as it took to make the dive—long enough for the body to eliminate the nitrogen bubbles from the blood. (See http://www.skin-diver.com/departments/scubamed/FreedivingCauseDCS.asp.)

−650

73 *most acute sense yet discovered on the planet:* So far, electroreception has been measured in sharks only when they are in very close range, about three feet from their targets. Researchers believe that sharks use it to orient their jaws for an accurate final attack. For instance, in the last few feet of an attack, great white sharks have been documented rolling their eyes back into their heads for protection and letting their electroreceptive sense guide them. See http://science.howstuffworks.com/zoology/marine-life/electroreception1.htm.

75 *speaker of Tzeltal:* Deutscher, *Through the Language Glass.*

77 *significantly more accurately than those with magnets:* Of the control-group students (the ones without the magnets), 77 percent pointed toward the home direction with 75 percent accuracy, but only 50 percent of the magnet-wearing students pointed accurately. Additional tests yielded similar results. See Baker, *Human Navigation,* 52.

77 *one in two hundred:* About ten years ago, researchers at the University of Western Ontario began a battery of tests on the effects of very low magnetic fields in the brain. The data from the majority of these tests showed that very low magnetic fields had a consistent and sometimes profound effect on the areas of the brain that processed unconscious thoughts and senses. In one test, conducted in 2009, researchers looked at the effects of very low magnetic fields on thirty-one volunteers. This experiment was designed to find the exact areas and mechanisms in the brain that were affected by the low magnetic fields. To do that, researchers put each volunteer in an fMRI scanner and prodded him or her with a heated stick. Next, the researchers broke the volunteers into two groups. In the control group, they repeated the exact same experiment with no changes; in the other group, they exposed each volunteer to a magnetic field of

no more than two hundred microteslas (a measurement of magnetic flux).

Neither the control group nor the exposed group reported feeling a difference between the two tests in the level of pain inflicted by the heated rod. However, the scans of the exposed group showed significant changes in the areas of the brain associated with processing pain (anterior cingulate, insula, hippocampus). The brains of the group exposed to the low magnetic field processed fewer pain signals, although the members of this group hadn't consciously known it.

The results of the study suggest that the effects of low magnetic fields aren't patent — that is, consciously sensed — but latent. In other words, they could affect and influence brain function without us realizing it.

The Earth's magnetic field ranges between twenty-five and sixty-five microteslas, about four times less than the field used in the experiment. Whether the subtle magnetic field of the Earth would be strong enough to make the human brain sense direction, though, nobody knows. However, the results of the experiment were significant enough to provoke the Western Ontario researchers to comment: "Magnetoreception may be more common than presently thought."

## −800

83 *French world champion freediver Audrey Mestre:* Many blamed Mestre's death on husband Ferreras, who was responsible for filling the air tank on the sled. According to a close friend and freediver partner, Carlos Serra, Ferreras was jealous of his wife's success and the two were on the brink of divorce. Serra and others speculated that Ferreras might have intentionally left the tank empty. Some crew members even recalled asking Ferreras repeatedly before the dive if he'd filled the tank, to which Ferreras repeatedly answered that he had. Even today, many freedivers put much of the blame for Mestre's death on Ferreras. Ferreras has always maintained his innocence. Authorities in the Dominican Republic absolved him of any crime. A year after Mestre's death, Ferreras made his own no-limits dive to 561 feet.

85 *dive to depths below 2,400 feet:* Irving learned Weddell seals were so adapted to diving that they seemed to gain oxygen from the water the deeper they dove. "From these various accounts, we can con-

clude that certain mammals have a capacity to resist asphyxia which far exceed the ability of man, and, as a matter of fact, exceed the capacity which we would expect on the basis of oxygen stored."

90 *Some claimed the ama could stay underwater for fifteen minutes at a time:* Later scientific investigation in the twentieth century only deepened the mystery. When Dr. Gito Terouka, a director of Japan's Institute for Science of Labour, came to monitor working conditions in Japan's southeastern coast he was dumbfounded by the ama's diving ability. He watched as an ama dove to depths below eighty-five feet for two minutes at a time. Even in winter, each ama wore only a thin cotton skirt, though the seawater crept below 50 degrees. To Terouka, a medical doctor by training, it seemed impossible. The pressure at eighty-five feet down was about two-and-a-half times that of the surface, enough to crush human organs and collapse the lungs. Further, the ama should have been suffering from hypothermia within the first hour. And yet they weren't. For hours, every day, for decades, the ama dove to extreme depths in frigid temperatures and enjoyed, for the most part, perfectly good health. Some ama even dove into their seventies and eighties. Terouka conducted a number of tests on the ama. He probed them, prodded them, and measured their inhalations and exhalations before and after deep dives, looking for some kind of clue to their apparent amphibious powers. His paper "Die Ama und ihre Arbeit," published in German in 1932, was the first-ever scientific review of breath-hold diving. It offered more questions than answers. The ama's myth only grew.

In the 1940s the Nazis, inspired by Terouka's work, conducted tests of their own on the human body's adaptability underwater. Replicating the ama's daily dive schedule, the Nazis plunged naked victims into ice-cold water for hours and monitored the molecular, physiological, and behavioral changes that took place. They tested recovery time by throwing victims directly from the ice-cold water into boiling water, exposing hypothermic victims to extreme heat, and injecting them with serums. They deprived victims of oxygen until they passed out, had them breathe mixed gases, carbon dioxide, and more. Most of the data from these grotesque experiments was later destroyed. The little that remained was deemed inconclusive.

Perhaps the real discovery was what Terouka and the Nazis didn't find. They didn't find the ama themselves were special in any way beyond having slightly larger lungs than the average woman and a little more fat to insulate against the cold water. They didn't find any genetic aberration or amphibious trait in their bodies. It was something else entirely, a secret that even today scientists are just at the cusp of truly understanding.

## −1,000

103  *other cetaceans have x-ray vision:* In the early 1940s, Arthur McBride, the curator of a marine park in St. Augustine, Florida, heard about this research and began to suspect that dolphins, whose behavior he observed extensively as part of his job, might echolocate as well. He kept a detailed record of his observations for a decade, but he died in 1950 before he could categorically prove that echolocation existed among dolphins.

Winthrop Kellogg, an American psychologist, continued McBride's work. He gathered two dolphins and placed them in a pool. In the middle of the pool, Kellogg had installed a large net with a hole at each end just large enough for a dolphin to fit through. Kellogg watched as the dolphins easily located the holes and briskly swam back and forth through them. At random, he closed off one of the holes with a piece of clear Plexiglas, which was invisible underwater. Kellogg moved the Plexiglas from one hole to the next. The dolphins had no idea which hole would be covered; underwater, the two holes also looked exactly the same. Yet they chose the Plexiglas-free opening ninety-eight out of a hundred times.

To Kellogg, these experiments proved that dolphins were using something other than vision to navigate, and it was probably echolocation. But others wondered if dolphins just had really good eyesight and could see a reflection of the Plexiglas underwater. In 1960, Kenneth Norris, a zoologist from the University of California, Los Angeles, proved dolphin echolocation once and for all.

He built a maze of vertical pipes in a swimming pool, each separated by only a few feet. Next, he brought in a dolphin and stuck rubber suction cups over its eyes. The cups were an effective blindfold, completely cutting off the animal's vision. Norris released the temporarily blinded dolphin into the swimming pool. It shot through the water, nimbly avoiding the pipes. Norris then threw a fish in the maze. The dolphin immediately swam between the pipes, located the fish, and ate it. During the fifty-eight experiments, the blinded dolphin did not collide with a single pipe. Norris had proved not only the existence but the remarkable accuracy of dolphin echo-location.

113 *equivalent of a human talking on the phone while chatting online:* Lilly said of the experiment, "They can be talking with whistles and talking with click trains, the whistles and the clicks completely out of phase with one another. They can be using the silence of the whistle exchange with the click exchange and filling the silences of click exchange with the whistle exchange, in this each are polite in their same mode. Thus one pair of dolphins talking can sound like two pairs of dolphins talking, one pair exchanging clicking, the other pair exchanging whistles." (See Lilly and Miller, "Vocal Exchanges Between Dolphins.")

117 *cracking the cetacean language code in the next few years:* As Kuczaj and I nibble on croissants and wait for the rest of the crew to arrive, he fills me in on more of the star-crossed history of dolphin-language research. Around the time Lilly opened CRI, the U.S. Navy hired three scientists to work with the U.S. Naval Undersea Warfare Center to create a machine that could translate human speech to dolphin whistles and back again. They named it the Man/Dolphin Communication Project. By 1964, the project's team, led by Dr. Dwight Wayne Batteau, a professor of physics and mechanical engineering at Harvard, was running trials on two dolphins, Puka and Maui, inside a secret laboratory in Hawaii. The machine, called a man-dolphin translator, worked like this: Batteau would speak an English word into a mike and the audio signal from the microphone would translate the word into a matching dolphin whistle, which would be broadcast through an underwater speaker into a pool outside the laboratory. When Puka and Maui responded, the translator would work in reverse, processing the whistle into a matching English word.

Patrick Flanagan, one of the scientists working with Batteau, claimed that the man-dolphin translator could successfully process thirty-five shared words between dolphins and humans. With the learned vocabulary, Puka and Maui were able to create simple sentences and respond to questions. Flanagan predicted that, within ten years, the team would establish a five-hundred-word shared vocabulary. In 1967, the team finished the research and were preparing a final report. They wrote that the project had successfully established verbal communication between humans and dolphins, and Batteau insisted that the project continue and the shared vocabulary be expanded. His claim made national news, and Harvard invited Batteau to give a lecture on his research. Unfortunately, before the final report was filed, he was found dead at a beach outside his house. The coroner's report said the cause of death was asphyxia due to drowning. To some, it was suspicious; Batteau was an excellent swimmer and had been in perfect health. The U.S. Naval Undersea Warfare Center shut down the Man/Dolphin Communication Project and classified all records. Most records of the projects have since disappeared.

Patrick Flanagan, who in 1961 was recognized by *Life* magazine as "one of the 100 most important young men and women in the United States," went on to study the mystical power of pyramids. He now sells face lotion and a water additive called Crystal Energy that he claims turns tap water into an elixir that can promote health and longevity. His YouTube videos garner hundreds of thousands of views.

By the 1980s, two scientists at the Russian Academy of Sciences in Moscow claimed to have identified more than three hundred thousand units of communication shared by dolphins. In one report, the lead scientist, Vladimir I. Markov, wrote that dolphins exchanged information through a wide range of acoustic signals, similar to a tonal language like Cantonese. These signals were organized like human language and included phonemes, which the dolphins combined into syllables, then words, and finally full sentences. Dolphins, Markov reported, had an alphabet consisting of fifty-one pulsed sounds and nine natural tonal whistles. His 1990 paper "Organization of the Communication System in *Tursiops truncatus montagu*" was published to little fanfare. The next year, the Soviet Union dissolved, and funding for Markov's communication projects dried up. Markov himself went off the map.

In the past three decades, research into dolphin communication has largely been co-opted by the New Age movement. Websites make spurious claims that dolphin echolocation can heal chronic depression, reverse Down syndrome, and correct various degenerative diseases. Dolphin-guided swims have become a multimillion-dollar business, even as dolphin communication research has become a fringe science.

The few legitimate researchers in the field are often forced to work outside academic or government institutions, raising their own money. Not many of them bother. One who has is Dr. Denise Herzing, a marine biologist who's been studying dolphins for two decades. For the past fifteen years, she has spent six months annually in the Bahamas trying to create a new dolphin-English transla-tor. In 2011, she enlisted the help of artificial intelligence engineers at Georgia Tech. The prototype for the system, called Cetacean Hearing and Telemetry (CHAT), has so far failed in all its lab and field trials.

121 *neither he nor Kozak ever tested the hypothesis:* In 2010, Jack Kassewitz, a freelance dolphin scientist in Florida, claimed to have proved holographic communication by recording a dolphin echo-locating against a triangular object and then playing the recording back to another dolphin, which immediately recognized the signal and fetched the triangular object from the seafloor. Kassewitz has yet to submit any details of his experiment to the scientific commu-nity. One researcher I talked to dismissed him as a "well-intentioned New Age dreamer."

## −2,500

156 *live below three thousand feet:* Nobody can agree on the actual num-bers of undiscovered species in the ocean — or on land — because no-body knows. The numbers I used here were gathered from *The Deep,* by Claire Nouvian, and 2012 presentation slides provided by Bruce Robison, research division chair at Monterey Bay Aquarium Research Institute. A 2011 study published in *PLOS Biology* (http://www.plosbiology.org/article/info:doi/10.1371/journal.pbio.1001127) sug-gests that there are only about 500,000 to 1,000,000 undiscovered species in the ocean, not including viruses and bacteria (which are nearly impossible to count). The Census of Marine Life Scientists (http://www.sciencedaily.com/releases/2011/08/110823180459.htm) argues that there are 6.5 million species on land (86 percent of which are undiscovered) and the percentage of these species at depths below three thousand feet is not known, mostly because humans have inves-tigated less than 1 percent of this area. On average, 50 to 90 percent of the specimens pulled up from nets at depths below three thousand feet are unidentified specimens, and, in most cases, new to science.

## −10,000

172 *surprising similarities between the two species:* But size isn't every-thing. There are multiple factors that affect intelligence, includ-

ing complexity of the cortex and presence of particular brain cells, such as spindle cells. Because most of an animal's brain is relegated to overseeing bodily functions, scientists in the 1960s argued that a more accurate gauge of intelligence is the proportion of brain to body mass. The thinking here is that the more body mass an animal has, the more brain it will need for bodily functions; any excess of brain mass would then most likely be used for higher-level thinking and thus could suggest higher intelligence. The calculation they created to compare brain to body mass was called the encephalization quotient, or EQ. An EQ of 1 means the animal has an average amount of brain for the body mass it is controlling. Humans have the highest EQ, around 7. That makes the human brain about seven times larger than would be expected given body size. Our cousins the chimpanzees have an EQ of about 2.5. Dogs don't fare as well, with an EQ of 1.7. Cats weigh in at an EQ of 1 — exactly average. While the bottlenosed dolphin scores an impressive 4.2 (the second highest of all animals), the sperm whale comes in at a dismal 0.3 — about 30 percent of what you'd expect to find in an animal of its size. The same EQ as a rabbit.

However, more recent research suggests that the EQ is an inferior measure of gauging potential intelligence compared to whole brain size and the ways in which an animal's brain evolved. Critics of the EQ point to animals such as the whale shark, which can grow up to forty feet long and weigh as much as forty-seven thousand pounds and yet has a brain the size of 36 grams, giving it an EQ of just .45. And then there are birds, which have extremely small brains but have demonstrated remarkable cognitive functions, including communication and tool use. Other animals, such as jellyfish, have no brain at all but know how to hunt, mate, and function in environments under extreme stress.

When I discussed the merits of the EQ with Stan Kuczaj, he summed it up by saying: "We simply don't know how the brain works well enough to make any of these assumptions with any of these equations at this time."

172  *"these are extremely intelligent animals"*: http://www.newscientist. com/article/dn10661-whales-boast-the-brain-cells-that-make-us-human.html.

176  *sperm whale language may be digital:* While creaks are used strictly for short-range echolocation, usually within an area of a few thousand feet, regular echolocation and social clicks can extend for tens or hundreds of miles. Low-frequency, slow sperm whale clicks may

even carry from one side of the planet to the other along something called the SOFAR (sound fixing and ranging channel) – a depth of two thousand to four thousand feet where sound can travel great distances without dissipating. It's the same basic effect as speaking into a tin can connected by a string to another can.

In the 1950s, the U.S. Navy sunk hydrophones down to the SOFAR channel to listen in on distant enemy submarines. Along with hearing sub noises, engineers started picking up strange moaning sounds, which they named the Jezebel Monster, after the name of the top-secret sub-surveillance project they were running. It wasn't a monster at all but the vocalizations of blue and fin whales. Whales, it appeared, had been using the SOFAR channel to keep in contact with one another hundreds, perhaps thousands, of miles apart.

Later, in the 1990s, an international group of scientists collaborated to construct a giant telescope and sink it eight thousand feet below the ocean's surface, off the coast of Toulon, France. The telescope, called Antares, was designed to detect the neutrinos, subatomic particles that the scientists believed could help them understand black holes and dark matter. When Antares was deployed in 2008, the first things it picked up weren't neutrinos but whale songs. It turned out that whales had evolved to transmit vocalizations across the deep ocean at the same superefficient wavelength that enables subatomic particles to traverse millions of miles through deep space.

179   *the whale's body on a south-facing beach:* For what it's worth (probably not much to most of you), the events of the Hussey legend, now centuries old, don't jibe with historical fact. Historians argue that, per the records, at that time, "Christopher Hussy" would have been either a six-year-old boy or twenty years deceased. They suspect the Hussey of legend was probably one of Christopher's grandsons. But nobody really knows.

181   *about 20 percent of the total population:* Ellis, *The Great Sperm Whale.*

183   *cetacean research feels like a race against time:* These figures taken from the 2005 documentary *A Life Among Whales* (IndiePix Films, 2009).

# Bibliography

Anderson, Kelly. "Inside Windfall Films' 'Sperm Whale.'" Realscreen.
  com, August 5, 2011. http://realscreen.com/2011/08/05/inside-
  windfall-films-sperm-whale/.
Ashcroft, Frances. *Life at the Extremes: The Science of Survival.* Berkeley:
  University of California Press, 2000.
———. *The Spark of Life: Electricity in the Human Body.* New York: W. W.
  Norton, 2012.
Baker, Robin. "Human Navigation and Magnetoreception: The
  Manchester Experiments Do Replicate." *Animal Behaviour* 35, no. 3
  (1987): 691–704.
———. *Human Navigation and the Sixth Sense.* New York: Simon and
  Schuster, 1981.
Bartle, Elinor. "The Secrets of the Deep." *Mar-Eco.* Accessed 2013.
  http://www.mar-eco.no/learning-zone/backgrounders/chemistry/
  The_Secrets_of_the_Deep.
Begley, Sarah. "The Deepest Dive." *TheDailyBeast.com*, July 23, 2013.
  http://www.thedailybeast.com/witw/articles/2013/07/23/no-limits-
  espn-s-nine-for-ix-explores-the-tragic-tale-of-freediver-audrey-
  mestre.html.
Bert, Paul. *Barometric Pressure: Researches in Experimental Physiology.*
  Durham, NC: Undersea Medical Society, 1978.
Bevan, John. *The Infernal Diver: The Lives of John and Charles Deane,*

*Their Invention of the Diving Helmet, and Its First Application.*
Hampshire, UK: Submex Ltd., 1996.

Boyle, Rebecca. "Divers Attempt to Communicate with Dolphins Using a Two-Way Translation Device." *Popular Science*, May 9, 2011. http://www.popsci.com/science/article/2011-05/dolphin-rosetta-stone-could-enable-two-way-communication-between-dolphins-and-humans.

Braconnier, Deborah. "Sperm Whales Have Individual Personalities." *PhysOrg.com*, March 16, 2011. http://phys.org/news/2011-03-sperm-whales-individual-personalities.html.

Branch, John, Adam Skolnick, William Broad, and Mary Pillon. "A Diver's Rise, and Swift Death, at the Limits of a Growing Sport." *New York Times*, November 18, 2013.

Broad, William J. *The Universe Below: Discovering the Secrets of the Deep Sea.* New York: Touchstone, 1997.

Bryner, Jeanna. "Dolphins 'Talk' Like Humans, New Study Suggests." *Livescience.com*, September 7, 2011. http://www.livescience.com/15928-dolphins-whistles-talk-humans.html.

Bulbeck, Chilla. *Facing the Wild: Ecotourism, Conservation, and Animal Encounters.* New York: Routledge, 2004.

"Bull Shark (*Carcharhinus leucas*)." *Arkive.* http://www.arkive.org/bull-shark/carcharhinus-leucas/.

Clapham, Philip. "Mr. Melville's Whale." *AmericanScientist.org* (book review), 2011. http://www.americanscientist.org/bookshelf/pub/mr-melvilles-whale.

Connor, Steve. "A Million Species of Animals and Plants Live in the Ocean Say Scientists." *Independent.co.uk*, November 15, 2012. http://www.independent.co.uk/news/science/a-million-species-of-animals-and-plants-live-in-the-ocean-say-scientists-8320295.html.

Cranford, Ted. "Faculty Profile." San Diego State University – Biology. Accessed 2013. http://www.spermwhale.org/SDSU/cranford.html.

Cromie, William. "Meditation Changes Temperatures: Mind Controls Body in Extreme Experiments." *Harvard University Gazette Archives*, 2002. http://news.harvard.edu/gazette/2002/04.18/09-tummo.html.

Deutscher, Guy. *Through the Language Glass: Why the World Looks Different in Other Languages.* New York: Picador, 2011.

*Discovery News* article, quoting journal *Current Biology*/Marine Register people: http://news.discovery.com/animals/whales-dolphins/marine-species-unknown-121115.htm.

Dolin, Eric J. *Leviathan: The History of Whaling in America.* New York: W. W. Norton, 2008.

"The Dominica Sperm Whale Project." *thespermwhaleproject.org.* Accessed 2013. http://www.thespermwhaleproject.org/.

Downey, Greg. "Getting Around by Sound: Human Echolocation." *PLOS Blogs: Neuroanthropology*, June 14, 2011. http://blogs.plos.org /neuroanthropology/2011/06/14/getting-around-by-sound-human-echolocation/.

Ellard, Colin. *You Are Here: Why We Can Find Our Way to the Moon but Get Lost in the Mall.* New York: Doubleday, 2009.

Ellis, Richard. *The Great Sperm Whale: A Natural History of the Ocean's Most Magnificent and Mysterious Creature.* St. Lawrence: University Press of Kansas, 2011.

Elsner, Robert. "Cardiovascular Defense Against Asphyxia." *Science* 153, no. 3739 (1966): 941–49.

———. "Circulatory Responses to Asphyxia in Pregnant and Fetal Animals: A Comparative Study of Weddell Seals and Sheep." *Yale Journal of Biology and Medicine* 42, nos. 3/4 (1969): 202–17.

———, and Brett Gooden. *Diving and Asphyxia.* New York: Cambridge University Press, 1983.

"Embryos Show All Animals Share Ancient Genes." *Discovery News*, 2013. http://news.discovery.com/animals/ancient-genes-embryos.html.

Ferretti, Guido. "Extreme Human Breath-Hold Diving." *European Journal of Applied Physiology* 84, no. 4 (2001): 254–71.

Finkel, Michael. "The Blind Man Who Taught Himself to See." *Mensjournal.com*, March 2011. http://www.mensjournal.com/ magazine/the-blind-man-who-taught-himself-to-see-20120504.

Gambino, Megan. "A Coral Reef's Mass Spawning." *Smithsonian.com*, 2009. http://www.smithsonianmag.com/arts-culture/A-Coral-Reefs-Mass-Spawning.html#ixzz1sEz3mD7z.

"Giant Amoebas Discovered 6 Miles Deep." *CBS News—Our Amazing Planet*, 2011. http://www.cbsnews.com/8301-205_162-20124830/ giant-amoebas-discovered-6-miles-deep/.

Goldenberg, Suzanne. "Planet Earth Is Home to 8.7 Million Species, Scientists Estimate." *TheGuardian.com*, August 23, 2011. http://www. theguardian.com/environment/2011/aug/23/species-earth-estimate-scientists.

Gregg, Justin. "Dolphins Aren't As Smart As You Think." *Wall Street Journal Online—Life and Culture*, December 18, 2013. http://online. wsj.com/news/articles/SB10001424052702304866904579266183573 854204.

Hagmann, Michael. "Profile: Gunter Wachterhauser Between a Rock and a Hard Place." *Science* 295 (2002): 2006–07. http://www.nytimes

.com/1997/04/22/science/amateur-shakes-up-ideas-on-recipe-for-life.html?pagewanted=all&src=pm.

Hansford, Dave. "Moonlight Triggers Mass Coral 'Romance.'" *National Geographic News*, 2007. http://news.nationalgeographic.com/news/2007/10/071019-coral-spawning.html.

Herman, L. M. "Seeing Through Sound: Dolphins (Tursiops Truncatus) Perceive the Spatial Structure of Objects Through Echolocation." *Journal of Comparative Psychology* 112, no. 3 (1998): 292–305.

Hoare, Philip. "The Cultural Life of Whales." *TheGuardian.com*, 2010. http://www.guardian.co.uk/science/2011/jan/30/whales-philip-hoare-hal-whitehead.

——. *Leviathan or The Whale*. London: Fourth Estate, 2008.

Hughes, Howard C. *Sensory Exotica*. Cambridge, MA: MIT Press, 1999.

"Humans and Gills." *Ask a Scientist!* DOE Office of Science, 2012. http://www.newton.dep.anl.gov/askasci/bio99/bio99850.htm.

Irving, Laurence, and Scholander, P. F. "The Regulation of Arterial Blood Pressure in the Seal During Diving." *American Journal of Physiology* 135, no. 3 (1942): 557–66.

Johnsen, Sonke, and Kenneth Lohmann. "Magnetoreception in Animals." *Physics Today* (March 2008): 29–35.

Kaharl, Victoria. *Water Baby*. Oxford: Oxford University Press, 1990.

Kemp, Christopher. *Floating Gold: A Natural (and Unnatural) History of Ambergris*. Chicago: University of Chicago Press, 2012.

Klimley, Pete. "Electroreception in Fishes: The Sixth Sense." *Biotelemetry* UC Davis. Accessed 2013. http://biotelemetry.ucdavis.edu/papers/WFC121_Electroreception.pdf.

Lang, T. G., and H.A.P. Smith. "Communication Between Dolphins in Separate Tanks by Way of an Electronic Acoustic Link." *Science* 150, no. 3705 (1965): 1839–44.

Langdon, J. H. "Umbrella Hypothesis and Parsimony in Human Evolution: A Critique of the Aquatic Ape Hypothesis." *Journal of Human Evolution* 33, no. 4 (1997): 479–94.

Layton, Julia. "How Does the Body Make Electricity—and How Does It Use It?" *Science.howstuffworks.com*. Accessed 2013. http://science.howstuffworks.com/life/human-biology/human-body-make-electricity1.htm.

Lilly, J. C., and A. M. Miller. "Vocal Exchanges Between Dolphins." *Science* 134 (1961): 1873–76.

Lilly, John C. *Communication Between Man and Dolphin: The Possibilities of Talking with Other Species*. New York: Crown, 1978.

——. "Critical Brain Size and Language." *Perspectives in Biology and Medicine* 6 (1963): 246–55.

——. *Man and Dolphin.* New York: Doubleday, 1961.

——. *The Mind of the Dolphin: A Nonhuman Intelligence.* New York: Doubleday, 1967.

Lindholm, Peter, and Claes E. G. Lundgren. "The Physiology and Pathophysiology of Human Breath-Hold Diving." *Journal of Applied Physiology* 106 (2009): 284–92.

"The Living Sea." *Oceans Alive.* Accessed 2013. http://legacy.mos.org/oceans/life/index.html.

Martinez, Dolores. *Identity and Ritual in a Japanese Diving Village: The Making and Becoming of Person and Place.* Honolulu: University of Hawai'i Press, 2004.

Marx, Robert F. *Deep, Deeper, Deepest: Man's Exploration of the Sea.* Flagstaff: Best Publishing, 1998.

Matsen, Brad. *Descent: The Heroic Discovery of the Abyss.* New York: Vantage, 2005.

"Meet Jonathan Gordon." *Nature: Sperm Whales – the Real Moby Dick.* Accessed 2013. http://www.pbs.org/wnet/nature/spermwhales/html/gordon.html.

Milius, Susan. "Moonless Twilight May Cue Mass Spawning." *ScienceNews.org,* 2011. https://www.sciencenews.org/article/moonless-twilight-may-cue-mass-spawning.

Mind Matters. "Are Whales Smarter Than We Are?" *ScientificAmerican.com,* January 15, 2008. http://www.scientificamerican.com/blog/post.cfm?id=are-whales-smarter-than-we-are.

"Moby Dick's Boom Box: Sound Production in Sperm Whales." *Ocean Portal, Smithsonian Museum of Natural History* (video on website), 2013. http://ocean.si.edu/ocean-videos/moby-dicks-boom-box-sound-production-sperm-whales.

Mora, Camilo, and Derek P. Tittensor. "How Many Species Are There on Earth and in the Ocean?" *PLOSBiology,* August 23, 2011. http://www.plosbiology.org/article/info:doi/10.1371/journal.pbio.1001127.

Morelle, Rebecca. "'Supergiant' Crustacean Found in Deepest Ocean." *BBC News Science and Environment,* February 2, 2012. http://www.bbc.co.uk/news/science-environment-16834913.

Morgan, Elaine. "Elaine Morgan: I Believe We Evolved from Aquatic Apes." *TED.com* (TED Talk video), 2009. http://www.ted.com/talks/elaine_morgan_says_we_evolved_from_aquatic_apes.html.

Morgan, Kendall. "A Rocky Start: Fresh Take on Life's Oldest Theory." *Science News* 163, no. 17 (April 26, 2003): 264.

Mueller, Ron, and Arek Piątek. "Beyond the Possible: Herbert Nitsch." *Red Bulletin*, March 5, 2013. http://www.redbull.com/cs/Satellite/en_US/Article/Freediver-Herbert-Nitsch-featured-in-April-2013-Red-Bulletin-magazine-021243322097978.

"Muscular Problems in Children with Neonatal Diabetes Are Neurological, Study Finds." *Science Daily Science News*, July 4, 2010. http://www.sciencedaily.com/releases/2010/07/100701145525.htm.

Nouvian, Claire. *The Deep*. Chicago: University of Chicago Press, 2007.

Ocean Register. November 2012. http://www.independent.co.uk/news/science/a-million-species-of-animals-and-plants-live-in-the-ocean-say-scientists-8320295.html.

O'Hanlon, Larry. "Giant Whale-Eating Whale Found." *Discovery News – Dinosaurs*, June 30, 2010. http://news.discovery.com/animals/giant-whale-fossil.html.

Palmer, Jason. "Human Eye Protein Senses Earth's Magnetism." *BBC News Science and Environment*, June 2011. http://www.bbc.co.uk/news/science-environment-13809144.

Pellizari, Umberto, and Stefano Tovaglieri. *Manual of Freediving*. Naples, Italy: Idelson-Gnocchi, 2004

Peralta, Eyder. "Researchers Find That Dolphins Call Each Other by 'Name.'" *The Two-Way: Breaking News from NPR* (Blog), February 20, 2013. http://www.npr.org/blogs/thetwoway/2013/02/20/172538036/researchers-find-that-dolphins-call-each-other-by-name.

Prager, Ellen. *Chasing Science at Sea*. Chicago: University of Chicago Press, 2008.

———. *Sex, Drugs, and Sea Slime*. Chicago: University of Chicago Press, 2011.

Rahn, H., and Tetsuro Yokoyama. *Physiology of Breath-Hold Diving and the Ama of Japan*. Washington, DC: Office of Naval Research, 1965.

Ravillous, Kate. "Humans Can Learn to 'See' with Sound, Study Says." *Nationalgeographic.com*, July 6, 2009. http://news.nationalgeographic.com/news/pf/35464597.html.

Reynolds, V., and Machteld Roede. *Aquatic Ape: Fact or Fiction?: Proceedings from the Valkenburg Conference*. London: Souvenir Press, 1991.

Rich, Nathaniel. "Diving Deep into Danger." *New York Review of Books*, February 2013. http://www.nybooks.com/articles/archives/2013/feb/07/diving-deep-danger/?pagination=false.

Robertson, John A. "Low-Frequency Pulsed Electromagnetic Field Exposure Can Alter Neuroprocessing in Humans." *Journal of the Royal Society Interface* 7, no. 44 (2010): 467–73.

Rosenbaum, Martin. "A Hunt for the Mysterious Beasts of the Deep (audio podcast)." *NPR Books: All Things Considered Author Interviews,* February 21, 2010. http://www.npr.org/templates/story/story.php?storyId=123898001.

Schmidt-Nielsen, Knut. *A Biographical Memoir: Per Scholander, 1905–1980.* Washington, DC: National Academy of Sciences, 1987.

Scholander, P. F. "The Master Switch of Life." *Scientific American* 209, no. 6 (1963): 92–106.

Seedhouse, Erik. *Ocean Outpost: The Future of Humans Living Underwater.* New York: Springer Praxis Books, 2010.

Severinsen, Stig Avail. *Breatheology.* Naples, Italy: Idelson-Gnocchi, 2010.

Shaefer, K. E. "Pulmonary and Circulatory Adjustments Determining the Limits of Depths in Breath-Hold Diving." *Science* 162, no. 3857 (1969): 1020–23.

Shubin, Neil. *Your Inner Fish: A Journey into the 3.5-Billion-Year History of the Human Body.* Vantage: New York, 2008.

———. *"Your Inner Fish" Lecture.* University of Chicago. Accessed 2013. http://tiktaalik.uchicago.edu/downloads/YourInnerFishLecture.ppt.pdf.

Siebert, Charles. "Watching Whales Watching Us." *New York Times Magazine,* 2009.

Skolnick, Adam. "A Deep-Water Diver from Brooklyn Dies after Trying for a Record." *New York Times,* November 17, 2013.

Smith, Hugh M. "The Pearl Fisheries of Ceylon." *National Geographic* 23, no. 1 (1912): 173–94.

Staaf, Daana. "Whales & Squid: Three Million Battles a Day." *Science 2.0: Squid a Day,* July 21, 2013. http://www.science20.com/squid_day/whales_squid_three_million_battles_day-116823.

Stromberg, Joseph. "How Human Echolocation Allows People to See Without Using Their Eyes." *Smithsonianmag.com,* 2013. http://blogs.smithsonianmag.com/science/2013/08/how-human-echolocation-allows-people-to-see-without-using-their-eyes/.

Summers, Becky. "Science Gets a Grip on Wrinkly Fingers." *Nature.com,* January 9, 2013. http://www.nature.com/news/science-gets-a-grip-on-wrinkly-fingers-1.12175.

*3-D Human Body,* First American Edition. New York: DK Children, 2011.

Touroka, Gito. "Die Ama und ihre Arbeit." *Arbeitsphysiologie* 5 (1932): 239–51.

"Two-Thirds Marine Species Remain Unknown." *Discovery News*, December 13, 2012. http://news.discovery.com/animals/whales-dolphins/marine-species-unknown-121115.htm.

"Underwater Exploration Timeline." *University of Wisconsin Sea Grant Institute*, 2001. http://www.seagrant.wisc.edu/madisonjason11/timeline/index_4500BC.html.

Verhoeven, Daan. "Freediving: Breaching the Surface of the Body's Capabilities." *Guardian*, September 16, 2013.

Viegas, Jennifer. "Dolphins: Second-Smartest Animals?" *Discovery News*, 2010. http://news.discovery.com/animals/whales-dolphins/dolphins-smarter-brain-function.htm.

———. "Dolphins Talk Like Humans." *Discovery News*, September 6, 2011. http://news.discovery.com/animals/dolphin-talk-communication-humans-110906.html.

Wade, Nicholas. "Amateur Shakes Up Ideas on Recipe for Life." *New York Times*, April 22, 1997. http://www.nytimes.com/1997/04/22/science/amateur-shakes-up-ideas-on-recipe-for-life.html?pagewanted=all&src=pm.

———. "Experiment Backs Up Novel Theory on Origin of Life." *New York Times*, August 25, 2000. http://www.nytimes.com/2000/08/25/us/experiment-backs-novel-theory-on-origin-of-life.html.

Wagner, Eric. "The Sperm Whale's Deadly Call." *Smithsonianmag.com*, December 2011. http://www.smithsonianmag.com/science-nature/The-Sperm-Whales-Deadly-Call.html.

"The Water in You." *USGS Water Science School*. Last modified August 9, 2013. http://ga.water.usgs.gov/edu/propertyyou.html.

"Whale Shark Specialty Student Manual SPC 641." *Georgia Aquarium* (June 2013): 10.

Whitlow, W. L. *The Sonar of Dolphins*. New York: Springer, 1993.

Yong, Ed. "Humans Have a Magnetic Sensor in Our Eyes, but Can We Detect Magnetic Fields?" *DiscoverMagazine.com*, June 21, 2011. http://blogs.discovermagazine.com/notrocketscience/2011/06/21/humans-have-a-magnetic-sensor-in-our-eyes-but-can-we-see-magnetic-fields/#.Usy2S2RDvxq.

Yopak, K. E., and L. R. Frank. "Brain size and Brain Organization of the Whale Shark, Rhincodon typus, Using Magnetic Resonance Imaging." *Brain, Behavior, and Evolution* 74, no. 2 (2009): 121–42.

# Index

Aborigines, 75
acoustic recording devices, 65–66,
    67–68, 102, 121–22, 189,
    221–22, 223–24. *See also*
    echolocation; tracking
*Alvin,* 139–40, 208
ama divers, 89–92, 94–97,
    244–45
amphibious reflexes, 5, 6–7, 29–30,
    128, 240–41
ancient cultures, 5, 75–76, 89–92,
    94–97, 244–45
*Angus,* 208
animals. *See also* marine life
    communication with, 109–16,
    117
anoxia, 129
apnea, 33, 130
Aqua Tek, 225–26
Aquarius, 12–18, 22–25, 226
Ashcroft, Frances, 153
Asia, freediving history of, 89–90,
    93

asphyxiation. *See* blackouts;
    breath-holding
astronauts, 146

Baker, Robin, 75, 76–77
Barclay, David, 211, 221–22
Bartlett, Doug, 199–201, 202, 204,
    211, 221
Barton, Otis, 60–61
bathypelagic zone, 139–41, 147–56,
    248
bathysphere, 59–61, 210
Batteau, Dwight Wayne, 246–47
Beebe, William, 59–61
Bert, Paul, 240–41
bioelectricity, 8, 153–54
bioluminescence, 148–49, 155–56
blackouts
    by King, 47–48, 226
    by Kitahama, 35
    by Mevoli, 228–29
    by Nestor, 159
    by Nitsch, 86–87

blackouts (*cont.*)
  physiology of, 40, 45, 128–29, 132
  by Pinon, 124–25
  safety, 33, 127–30
Blaine, David, 130
blindness, 8, 103–9
blood
  composition, 6
  decompression sickness and, 14,
    20, 44, 87–88, 241–42
  heart rate and, 28–30, 43
Boyle's law, 27
brain damage, 132. *See also*
  neurology
breath-holding, 38–41, 130–34,
  158–61
Bucher, Raimondo, 27
Buddhist monks, 8, 154
Bushway, Brian, 104–9
Buyle, Fred, 56–58, 63–72, 186, 225,
  229

Cameron, James, 210
cancer research, 148
carbon dioxide
  buildup, 39, 43, 158
  exchange, 147
Carpenter, Scott, 21
Cayman Trench, submarine tour of,
  143–56
cetaceans. *See* dolphins; whales
Challenger Deep, 210–12
chemosynthetic life, 209–10, 212–15
Click Research, 223–24
CNF. *See* constant weight without
  fins
communication
  dolphin, 8, 101–3, 109–16,
    118–22, 246–48
  holographic, 121–22, 224, 248
  human echolocation, 106–9
  human-dolphin, 109–16, 118–21,
    246–48

human-gorilla, 117
  "telepathic," of coral, 17, 240
  whale, 100–101, 111–12, 171–78,
    185–86, 190, 222–23, 249–50
competition
  Nestor's view on, 229–30
  no-limits, 84–88
  2011 Greece, 1–3, 30–36, 45–52
  2013 Vertical Blue, 227–29
competitions, freediving, 1–3,
  30–36, 45–52, 71, 130
compression. *See* pressure
computers, human correlation to,
  153n, 199
Condert, Charles, 19
Conshelf, 20–21
constant weight (CWT), 45, 162
constant weight without fins (CNF),
  34
construction, underwater, 19–20
convulsions, 39
Cook, James, 75
Corliss, Jack, 207–9
Coste, Carlos, 82
Cousteau, Jacques, 20–21
CWT. *See* constant weight
Czech freedivers, 50–52

da Vinci, Leonardo, 18–19
Dallenbach, Karl, 103–4
DareWin. *See* Schnöller, Fabrice
Deane, John, 19
decompression. *See also* blackouts
  equalization and, 44–45, 136, 137,
    160–61
  process, 42, 44–45
  by seals, 84–85
  sickness, 14, 20, 44, 87–88, 241–42
*DeepSea Challenger,* 210
Deep Sound, 211, 221–22
delirium, 23–24, 44, 129, 158
detritus, 147, 198–99, 204
Diderot, Denis, 103

diet, 129–30
diving apparatus, inventions, 18–20,
    59–61, 82–83, 210
dolphins, 8, 101–3, 109–17, 118–22,
    245–48
*Down and Out in Paris and London*
    (Orwell), 150

Eastern medicine, 154
echolocation
    dolphin, 8, 101–3, 109–12, 122, 245
    human, 103–9
    neurology of, 108, 122
    whale, 171–73, 187–88, 191–92,
        249–50
electrical charge
    of electric rays, 8, 152–53
    in humans, 8, 153–54
    magnetoreception and, 7–8,
        62–63, 72–78, 242–43
Elsner, Robert, 241
*Endeavour,* 75
equalization, 44–45, 136, 137,
    160–61
*Essex,* 180–81
euphoria, 23–24, 44, 129, 158
evolution, 74, 172, 200, 212–15
extrasensory abilities
    amphibious reflexes, 5, 6–7,
        29–30, 128, 240–41
    communication, of coral, 17, 240
    echolocation, 8, 101–12, 122,
        171–73, 187–88, 191–92
    holographic communication,
        121–22, 224, 248
    hypothermia resistance, 244–45
    magnetoreception, 7–8, 62–63,
        72–78, 242–43

*Falkor,* 202, 205–7, 215
Fernández de Oviedo, Gonzalo.
        *See* Oviedo, Gonzalo
        Fernández de

Ferreras-Rodriguez, Francisco, 44,
    83, 243
Fix, Markus, 66–69, 122, 224
Flanagan, Patrick, 246–47
FlashSonar, 106, 107–9
Forbes, Edward, 58
40 Fathom Grotto, 134–35
Franz, Benjamin, 82
*Free Fall,* 49
freediving
    amphibious reflexes for, 5, 6–7,
        29–30, 128
    breath-holding for, 38–41,
        130–34, 158–61
    certification, 131
    CNF, 34
    competition, 1–3, 30–36, 45–52,
        71, 130, 227–30
    CWT, 45, 162
    death by, 83, 227–29, 243
    decompression, 42, 44–45
    diet, 129–30
    disciplines, 34
    free immersion, 49–52, 228
    gravity in, 37, 43, 147, 194–96
    history of, 27, 89–90, 92–94,
        244–45
    hydrodynamics for, 135
    Japanese ama, 89–92, 94–97,
        244–45
    lung capacity for, 28–30, 38
    Nestor's experience, 135–38,
        193–97
    no-fins, 228–29
    no-limits, 82–88, 243
    pearl divers, 20, 92–93
    pressure in, 2–3, 7–9, 43–45
    Prinsloo's maxims of, 37, 164–65
    qualifications, 4
    research of sharks by, 57–58,
        66–69, 71–72, 78–79,
        225–26
    research of whales by, 187–88

freediving (*cont.*)
  as research tool, 9–11, 230
  safety, 127–30
  as spiritual practice, 3–4, 37, 49,
    89, 230
  static apnea, 130
  training, 135–38, 157–58
Frenzel method, 160–61

Galápagos, 71–72, 207–9
Gazzo, Guy, 66–69, 78–79, 191–92,
  231
genetics research, 148–49, 199–201,
  239–40
Ghislain, Jean-Marie, 164, 186,
  225–26
Godson, Lloyd, 15
gorillas, human communication
  with, 117
gravity, 37, 43, 147, 194–96
Greece, competitions in, 1–3, 30–36,
  45–52, 84–88
Guugu Yimithirr, 75

hadal zone, 199, 200–1, 203–5,
  207–16, 221–22
Halley, Edmond, 18
hallucinations, 44, 129, 131
Harty, Ted, 160–61
heart rate, in water, 28–30, 43
*Hercules,* 207–8
Herzing, Denise, 247–48
Holman, James, 103
holographic communication,
  121–22, 223, 248
Homer, 92
Howe, Margaret, 114–15
humans
  bioelectricity in, 8, 153–54
  computer correlation to, 153n, 199
  dolphin communication with,
    109–16, 118–21, 246–48
  echolocation by, 103–5

evolution of, 74, 172, 200, 212–15
intelligence of, animals versus,
  248–49
magnetoreception of, 74–78,
  242–43
ocean connection of, 6–7, 198–99,
  230, 239–40
underwater living for, 12–18,
  20–25, 226
whale communication and,
  175–76
Hussey, Christopher, 179, 250
hydrodynamics, 135
hydrophones, 102, 189
hydrothermal vents, 143, 209–10,
  212–15
hypothermia resistance, 244–45

*Idabel,* 140, 142–44
infants, amphibious reflex of, 6, 128
intelligence
  of dolphins, 111
  of humans, measuring, 248–49
  of whales, 101, 172–73
inventions
  acoustic tracking, 65–66
  diving apparatus, 18–20, 59–61,
    82–83, 210
  submarine, 139–41, 142–43
  underwater habitat, 20–22
Irving, Laurence, 84–85

Japanese ama divers, 89–92, 94–97,
  244–45

Kac, Eduardo, 148–49
Kanzi (gorilla), 117
Kassewitz, Jack, 248
Kellogg, Winthrop, 245
Kermadec Trench, 204
King, David, 45, 46–48, 226
Kish, Daniel, 106, 108
Kitahama, Junko, 34–35

Koko (gorilla), 117
Kozak, V. A., 121
Kuczaj, Stan
  dolphin tracking by, 116–17, 119
  submarine tour with, 142,
    143–56, 223

Leggat, Bill, 239–40
LEGGO, 215–16, 221–22
Lethbridge, John, 18–19
Lilly, John C., 112–14, 115–16, 246
Lucayans, 92–93
lung capacity, 28–30, 38. *See also*
  decompression

magnetoreception
  by animals, 7–8, 62–63, 72–74
  disruption technology, 225–26
  by humans, 74–78, 242–43
Malpelo island, 71–72
mammalian dive reflex, 5, 6–7,
  29–30, 128, 240–41
Marco Polo, 92
Mariana Trench, 20, 201–2, 210–12
marine life
  algae, 17
  in bathypelagic zone, 147–52,
    155–56, 248
  bioluminescent, 148–49, 155–56
  chemosynthetic, 209–10, 212–15
  color of, 145
  coral, 17–18, 239–40
  decompression techniques,
    84–85
  dolphins, 8, 101–3, 109–16,
    118–22, 245–48
  echolocation of, 8, 101–12, 122,
    171–73, 187–88, 191–92
  in hadal zone, 204–5, 208–10,
    212–16, 221–22
  Japanese ama dive for, 89–92,
    94–97
  jellyfish, 148–49, 155–56

  magnetoreception of, 7–8, 62–63,
    72–74
  in mesopelagic zone, 58–59, 62
  in photic zone, 6–17, 41, 145
  sea lily, 58–59
  seals, 29, 84–85, 135, 241, 243–44
  sharks, 7–8, 57–58, 64–74, 78–79,
    165–66, 225–26
  shrimp and shrimplike, 204, 209,
    216, 221
  snailfish, 204
  squids, 148, 155
  whales, 8, 100–101, 111–12,
    170–92, 222–23, 249–50
  xenophyophores, 204
Marine Mammal Protection Act,
  116
Markov, Vladimir I., 247
Marshall, Peter, 169, 232
Master Switch of Life, 5, 30, 43–44,
  196
Maui (dolphin), 246
McBride, Arthur, 245
meditation, 8, 154, 163
Merkel, Friedrich, 62–63
mesopelagic zone, 58–62, 145
Mestre, Audrey, 83, 243
Mevoli, Nicholas, 227–29
Mexico, ancient, 75–76
microbiology sampling, 199–201
midnight (bathypelagic) zone,
  139–41, 147–56, 248
Mifsud, Stéphane, 38, 130
Molchanova, Natalia, 82

National Oceanic and Atmospheric
  Administration (NOAA), 226
Navy, U.S., 21, 139, 246–47, 250
Nazis, 244
Néry, Guillaume, 49
Nestor, James
  ama divers visited by, 90–92,
    94–97

Nestor, James (*cont.*)
  Aquarius tour for, 22–25
  breath-hold training of, 38–41,
      131–34, 158–61
  competitive diving views of, 1–3,
      229–30
  dolphin tracking experience of,
      116–17, 118–22
  echolocation tried by, 107–9
  freedive training of, 135–38,
      157–58
  freedives by, 135–38, 193–97
  at Greece's World
      Championships, 1–3, 30–36,
      45–52
  at Mariana Trench, 210–12
  research methods of, 5–7, 9–11,
      179
  safety training of, 127–30
  shark fear of, 165–66
  shark-tracking experience of,
      66–69
  submarine deep-sea tour for,
      143–56, 223
  whale-tracking experience of,
      178–79, 183–86, 188
neurology
  of bioelectricity, 153–54
  of brain damage, 132
  of breath-holding, 132
  of dolphins, 111
  of echolocation, 108, 122
  of heart rate in water, 28–30
  of whales, 172–73
nitrogen narcosis. *See*
      decompression
Nitsch, Herbert, 80–82, 83–88, 224
Nixon, Richard, 116
NOAA. *See* National Oceanic and
      Atmospheric Administration
no-fins freediving, 228–29
no-limits freediving, 82–88, 243
Norris, Kenneth, 245

ocean. *See also* marine life
  bathypelagic zone of, 139–41,
      147–56, 248
  color of, 144–45
  detritus, 147, 198–99
  hadal zone of, 199, 200–1, 203–5,
      207–16, 221–22
  human connection to, 6–7,
      198–99, 230, 239–40
  mesopelagic zone of, 58–62, 145
  photic zone of, 16–17, 18, 41,
      145
  pollution in, 182–83
oil industry, 180–82
ooze, 198–99
Orwell, George, 150
Oviedo, Gonzalo Fernández de, 92
oxygen. *See also* blackouts
  breath-holding physiology and,
      39–41, 132
  Earth's exchange of, 17, 147
  gaining, in water, 28–30, 43,
      243–44
  static apnea with pure, 130
  toxicity, 84–85, 241

Payne, Roger, 182–83
pearl divers, 20, 92–93
Performance Freediving
      International, 125, 127
Peter (dolphin), 114–15
Petrović, Branko, 130
pharmaceuticals, 154
photic zone, 16–17, 18, 41, 145
photography, underwater, 56, 59,
      208
phytoplankton, 147, 198
Piccard, Jacques, 20, 210
Pinon, Eric
  blackout by, 124–25
  training by, 125–30, 131, 134,
      135–38
pollution, 182–83

pressure
  on Aquarius, 15, 24
  blood flow under, 14, 29–30, 40, 132
  Boyle's law of, 27
  delirium from, 23–24, 44
  equalizing, 44–45, 136, 137, 160–61
  in hadal zone, 204
  levels of, 7–9
  physiology and, 7–9, 13–14, 43–45, 194–96
prey
  of bathypelagic animals, 152
  of sharks, 73
  of whales, 222–23
Price, David 215
Prinsloo, Hanli
  freediving maxims of, 37, 164–65
  last competitive freedive of, 37, 161–63
  swimming with sharks, 165–66
  training from, 38–41, 163–64, 193–94
  whale tracking with, 178, 183–86, 192–93
Puka (dolphin), 246

reflexes, amphibious, 5, 6–7, 29–30, 128, 240–41
remotely operated underwater vehicles. See ROVs
Rendell, Luke, 187
rescue training, 128–30
Réunion Island, 55–56
  dolphin tracking off, 109–12, 116–17, 118–21
  shark tracking off, 57–58, 66–69, 71–72, 78–79, 225–26
Richet, Charles, 28
Rišian, Michal, 50–52
Roatan, 141–42, 223
Rosser, Saul, 12–13, 15–16

ROVs (remotely operated underwater vehicles), 139, 200, 206, 210–11
Russell, Michael, 214
Russian Academy of Sciences, 247
Rutten, Otto, 25–26
Ryumin, Valery, 146

safety training, 127–30
Sagan, Carl, 112–113, 114
sailors, 75, 179–81
sambas, 127–28
Sars, Michael, 58–59
Schmidt, Eric, 202
Schmidt, Wendy, 202, 206
Schnöller, Fabrice
  Click Research by, 223–24
  dolphin tracking by, 109–12, 116–17, 118–22
  shark tracking by, 57, 64–69, 78
  whale tracking by, 100–101, 173–74, 178–79, 186, 187, 188, 191–92, 222–23
Scholander, Per, 5, 29–30
scuba (self-contained underwater breathing apparatus), 19, 42
SEALAB II, 21n
seals, 29, 84–85, 135, 241, 243–44
seizures, 127–28, 129
senses. See extrasensory abilities
Serra, Carlos, 243
sharks
  attacks by, 56, 64–65, 225
  magnetoreception of, 7–8, 72–74
  swimming with, 165–66
  tracking, 57–58, 65–69, 71–72, 78–79, 225–26
Siestas, Tom, 130
Simon, Bob, 81
Sirena Deep, 201–2, 204
sonar. See acoustic recording devices; echolocation
space travel, 146

spiritual practice
  freediving as, 3–4, 37, 49, 89, 230
  meditation, 8, 154, 163
  yoga, 38–39, 163
Stanley, Karl, 140–41, 143–56, 223
static apnea records, 130. *See also*
  breath-holding
Štěpánek, Martin, 130
submarines
  inventions of, 10, 139–41, 142–43
  tours on, 143–56, 223
sunlight (photic) zone, 16–17, 18,
  41, 145

Takayan (Japanese guide), 90–91,
  94–95, 97
Terouka, Gita, 244
tracking
  dolphin, 109–12, 116–17, 118–22
  shark, 57–58, 65–69, 71–72,
    78–79, 225–26
  whale, 100–101, 168–69, 173–74,
    178–79, 183–92, 250
training
  breath-holding, 38–41, 131–34,
    158–61
  freediving, 135–38, 157–58
  safety, 127–30
*Trieste*, 20, 210
Trincomalee, Sri Lanka, whale
    tracking in, 168–69, 178–79,
    183–86, 189–92, 222–23
Trubridge, William, 2–3, 31–32, 34,
    35, 42–43, 48
Tsibliyev, Vasily, 146
Tupaia (Polynesian chief), 75

underwater construction, 19–20
underwater living, 12–18, 20–25,
  226

underwater photography, 56, 59,
  208

vagus nerve, 28
Valsalva maneuver, 160
Vaughan-Lee, Emmanuel, 189
Vertical Blue competition, 227–29
Vikings, 92

Wächtershäuser, Günter, 212–14
Walsh, Don, 20, 210
whales
  communication of, 100–101,
    111–12, 171–78, 185–86,
    190, 222–23
  echolocation by, 171–73, 187–88,
    191–92
  hunting, for oil, 180–82
  intelligence of, 101, 172–73, 182
  physiology of, 170, 172–73
  prey of, catching, 222–23
  tracking, 100–101, 168–69,
    173–74, 178–79, 183–92,
    222–23, 250
*Whales Weep Not*, 170, 181–82
Whitehead, Hal, 170–71, 187–88
Winram, William, 66–69
World Access for the Blind, 106,
  108
*A World Without Sun*, 21

yoga, 38–39, 163

zones
  bathypelagic, 139–41, 147–56,
    248
  hadal, 199, 200–1, 203–5, 207–16,
    221–22
  mesopelagic, 58–62, 145
  photic, 16–17, 18, 41, 145